A Brief History of Pollution

A Brief History of Pollution

ADAM MARKHAM

St. Martin's Press
New York

© Adam Markham, 1994

All rights reserved. For information, write:
Scholarly and Reference Division,
St. Martin's Press, 175 Fifth Avenue,
New York, NY 10010

First published in the United States of America in 1994

Printed and bound in Great Britain by
Clays Ltd, St Ives plc

ISBN 0-312-12368-X(Cloth)
ISBN 0-312-12369-8 (Paper)

Library of Congress Cataloging in Publication Data applied for.

To Victoria

To see a World in a Grain of Sand
And a Heaven in a Wild Flower
Hold Infinity in the palm of your hand
And Eternity in an hour.

William Blake

Contents

Acknowledgements

With the publication of this book I celebrate ten years of working in the environment movement. I should like to thank Jonathon Porritt for bringing me into the mainstream of environmental campaigning in 1984, and the collective staff of the magazine *Touch and Go/Cael a Cael* for nurturing my writing and political skills. Chris Rose led by example and encouraged me to break the rules. Gordon Shepherd tried to make me abide by them while keeping the bureaucrats at bay. Claude Martin gave me the chance to mould WWF's policies on pollution and consumption. Ivan Hattingh provided encouragement and contacts. To these few people, I owe much.

Inspiration to keep on writing and editing in the free moments away from my day job came from Vikram Seth, Peter Mathiessen, Homero Aridjis, Paula and William Merwin and John Arden. Malcolm Fergussen, Mark Barrett, Chris Rose, Rafe Pomerance, Dave Baldock, Nigel Haigh, Richard Mott, Nigel Dudley, Steve Elsworth, Marek Meyer, Scott Hajost, Gerard Peet, Rik Leemans, Mike Hulme, Christer Ågren, Claire Holman and Patrick Halpin have all helped in forming my views on the politics of pollution.

Michael Rae, Tessa Robertson, Konrad Meyer, Bill Eichbaum, Chris Hails, Pascale Mohrle, Chris Elliott, Anne-Marie Frusciante, Paolo Lombardi, Tom Mathew, Danny Elder, Tundi Agardy, Victoria Dompka, Dawood Ghaznavi, Martin Abraham, Paul Anthony, Bob Maude, Benôit Jacques and Gianfranco Bologna have given invaluable support for my work on pollution at WWF. Paul Hohnen, Bill Hare, John MacKinnon, Tim Whitmore, Tessa Tennant, Charles Clover, K V Ramani and

Betty Ferber de Aridjis should all share in some of the credit for this book. To all of these people, and the many more un-named individuals to whose writings and activities I owe a debt, I extend my thanks and admiration. I hope that in writing this short volume I have done some justice to their work and varied environmental causes. Despite my many debts to others in the development of this book, I must emphasize that any errors of fact or analysis remaining in the text are mine alone.

My parents, Bill and Priscilla Markham, as well as Ken Cook, Susan Sechler, and Tom and Mary Norman have been a constant source of support during the writing of this book. My wife, Victoria Dompka, is the one who made it possible for me to complete the project with her continual encouragement, advice and indulgence.

Adam Markham
29 March 1994

Preface

Let us not...flatter ourselves overmuch on account
of our human conquest over nature.
For each such conquest takes its revenge on us.

Frederick Engels

Our world has has been subjected to an unprecedented increase
in types and quantities of pollution during the last two hundred
years, and the onslaught grows worse day by day. Untrammelled
consumption in the industrialized nations is squeezing the world
dry of natural resources and unleashing a deadly torrent of waste
and contaminants. Struggling to overcome the debilitating
legacies of colonialism, fast-growing populations in the South
are set to try and emulate the consumption patterns of the North
with little regard for the environment. Meanwhile, the twin
economic deities of growth and production seem to blind the
human race to the existential necessity of remaining within the
carrying capacity of this fragile yet productive planet.

Throughout history human societies have gone forward in
the critically flawed belief that the oceans and rivers would wash
away filth, the winds cleanse the air and the soils bury the rest.
Instead the sheer quantity, and unrelenting flow, of human and
industrial waste have overwhelmed the natural assimilative
capacity of the earth and undermined human development. No
aspect of life on earth is untouched by the dread hand of pollution. Clean water, fresh air and pristine environments no longer
exist in anything more than concept. Unnatural chemicals such
as polychlorinated biphenyls (PCBs) now contaminate the
farthest reaches of wilderness, from the icy wastes of Antarctica
to the barren dryness of the Sahara. Pollution is ubiquitous,

seeping into and slowly rotting the fabric of the environment and affecting human societies and cultures in ways which we are only just beginning to fully comprehend.

Where were the signposts to this twentieth century tragedy? What exactly is pollution, and how did it come to dominate the environmental debate? What is the prognosis for the future? These are some of the questions I have tried to answer in this book. I have written neither an academic history of pollution nor a technical manual on how to control it. This is not a guide to saving the planet or even a doom-mongering catalogue of catastrophes. It does, however, provide the basis for any reader to engage in informed debate upon the topic, perhaps with some advantage.

What I have tried to do is provide a tourist guide to pollution, from prehistoric times to the post-industrial era, melding a discussion of the problems themselves with an attempt to understand the reasons for their emergence on a variety of political agendas. My intention has been to put pollution in a social and cultural context. For it is my belief that the environment movement has become too narrowly focused on the empirical demonstration of pollution damage, and on battles with polluters played out in the law courts, or more often, the mass media. In so doing, environmentalists have made themselves a political force to be reckoned with, while at the same time often becoming divorced from the realities of social need and economic development. I am a biologist and environmentalist not a historian and, therefore, no less susceptible to fault than others of the ilk. In attempting to redress the balance and link the problems of the present with the pressures of the past, I owe a huge debt to a few writers who have, I believe, pioneered the kind of history that allows us to put twentieth century problems in their true intergenerational context.

Fernand Braudel's spectacular histories of Spain, France and the World, along with Daniel Boorstin's works on the history of the USA seemed to me to pave the way. Stephen Jay Gould, with his extraordinary essays on science, sociology and palaeontology has blazed a trail worthy of Darwin, Engels or T H Huxley. And in their environmental histories, Clive Ponting, William Cronon, Peter Marshall and Keith Thomas have all performed a sterling service. Combining this kind of history with the testament of contemporary fiction, poetry and science gives the kind of cultural overview that is very hard to come by in today's environmental debate. I remain convinced that we will not solve our modern pollution problems without linking our scientific arguments to those of artists, writers and social scientists. Pollution is not a scientific problem, but a societal one with complex historical roots.

To give an example of how critical these links are with social history, it is worth noting that while writing this book I was struck by how deftly the human race has ignored the influence of consumerism on the rise of pollution. We have become more remote from nature and have centred our lives around the attainment of ever higher levels of material well-being. In this way we have ignored our individual impact on the environment and confused living standards with quality of life.

The answer does not lie in a back-to-the-land rush towards the not so idyllic rural ways of former centuries, but in a radical re-evaluation of our consumption patterns and needs. Until now, industry and business have been the twin bogeymen of conservation, yet we desperately need their cooperation, innovation and partnership if we are to reverse the trends. Business must move again towards the provision of society's needs and away from the primacy of profit. Production can no longer be the sole driving force of our world, nor consumption its primary end.

The 'population explosion' has been another favourite red rag to the bullish conservationists of the northern hemisphere. And although there is undeniably a need for population stabilization, it remains a sad fact that the combination of life expectancy and consumption patterns makes an American child 500 times as much of a burden on the environment as one born in Mali.

Equity demands that we of the north supress our resource hunger so that those in the south can meet their rising needs. Common sense dictates that people will not give up TVs, air-conditioning and frozen food, and that those who do not have these conveniences will continue to aspire to them. In this book I have tried to trace the roots of this problem and to locate them in a historical and cultural context. I have merely skimmed the surface of these arguments, but it is clear to me that until we trudge the foothills of our own over-consumption, we will never scale the heights of pollution prevention, much less conquer the peaks of equitable and sustainable resource use.

Adam Markham,
Washington DC
February 1994

Chapter One

A BRIEF HISTORY OF POLLUTION

Wherever the Dragon advances, the world darkens.
It belches clouds of soot and smoke, engulfing all
that is living in its burning breath.

Folke Isaksson

Soiling the Garden of Eden

Older than civilization, pollution has been a problem since the appearance of our earliest ancestors. The adulteration of water, soil and air by the physical and chemical waste products of human activity has accompanied our species since it first walked upon the earth. And from prehistoric times, pollution has been inextricably linked to health and medicine.

Palaeopathologist Paul Janssens[1] suggested that:

Medicine precedes religion. The preservation of life whether generally or individually fulfills an inborn need or instinct. It finds expression within the framework of medicine before any notion of religious feeling, since sickness has to exist before it can be looked upon as the punishment of an angry deity.

Much early sickness was undoubtedly caused by what we would today, call pollution. The very earliest form of pollution must have resulted from the act of defecation. The presence of human gut bacteria such as *Escherichia coli* in drinking water was the first water pollution and must have been a source of illness for prehistoric man, just as it is for millions of people today.

The discovery of fire, at least half a million years ago, created the first significant air pollution source, and smoke remains a major problem in the modern world. Ancient human communities are thought to have suffered from sinusitis and blackening of the lungs (anthracosis) due to regular exposure to smoke.[2]

Dust pollution also has early origins and Janssens speculated that the Neolithic miners of central Europe, who daily chipped flints from limestone quarries like that of Obourg, suffered from silicosis. Their every breath during the working day would have drawn in air polluted with dust from their labours.[3] Simple geography sometimes influenced historical exposure to pollutants. Recent analysis of the 200,000 year old Broken Hill hominid from Zambia has produced evidence that he suffered from lead poisoning due to an ore lode underlying the water supply of the cave dwelling.[4]

The transition from hunter-gathering to nomadic herding systems and eventually to settled agriculture during the Neolithic period has been described as 'the most fundamental change in human history.'[5] By allowing output of food to increase, the concept of 'property' to develop, and surplus food production to grow, the agricultural transition became the basis of a human revolution. Food surpluses enabled the development of non-farmers within society, including the priesthood, the army and craftsmen. The distribution and collection of food was the basis for power and the development of wealth, and the ability to produce more from a smaller area of land laid the basis for population growth.

Out of agriculture, grew the community. Small villages at first, then towns and eventually city-states. Jericho was a walled town of ten acres in 6500 BC, and the Mesopotamian temple city of Uruk had a population of 50,000 people by 3000 BC. For a modern comparison it is noteworthy that the French city of Toulouse had only reached a population of 55,000 nearly 5000 years later in 1789.[6] This development of towns and cities ushered in the pollution era.

There hasn't always been an encompassing word for the

filth, grime, miasma, smoke, slime, sludge and generally disagreeable and dangerous substances that contaminate our world. As late as 1783, Dr Johnson defined pollution as 'the act of defiling' or 'the contrary of consecration'. The verb, according to Johnson, meant 'to make unclean in the religious sense' or to 'taint with guilt.'[7] The use of the word pollution in its current sense only gained currency in the nineteenth century.

In 1972 the British biologist Kenneth Mellanby defined pollution as the 'presence of toxic materials introduced into our environment by man',[8] but it can also mean the disruption of natural soil and water regimes by the displacement or mobilization of natural substances. A classic modern example is the pollution of rivers and coastal ecosystems by soil and silt washed off the land due to deforestation or poor agricultural practices. Salinization is another phenomenon in this category, and it was this that destroyed Sumerian civilization.

Over a period of about 1700 years, from 3500 BC to 1800 BC, Sumarian agriculture declined and wheat productivity fell because of salt pollution. When flat land is irrigated, as it was in southern Mesopotamia, lack of proper drainage causes the water to seep into the underlying groundwater and raise the water table. As the soil becomes waterlogged, salts rise and high levels of evaporation from the soil surface leave salt crusted on the fields. Having invented writing, the Sumerians were able to record that the 'earth turned white'.[9]

According to the World Health Organization (WHO) and the United Nations Environment Programme (UNEP), even today 'increasing salinity is one of the most significant and certainly the most widespread forms of groundwater pollution'. Salinity now seriously affects 7 percent of the world's irrigated crop land, mainly in India (24 percent of the total irrigated area), the USA, Pakistan, Iran, Iraq and Egypt.[10] Thus a direct pollution lineage can be traced from ancient Sumer to the modern Middle East.

Unsanitary Conditions

The first sewage system was the Roman Cloaca Maxima, built in the sixth century BC during the Etruscan dynasty of the

Tarquins. The initial purpose of this massive structure was to drain the swamp between the Palatine and Capitoline hills, leading eventually to the establishment of the Roman Forum, which became the hub of the Republic and later the Empire. The hydraulic pioneers of the ancient world, the Romans constructed a whole network of *cloacae*, or sewers, as well as a maze of aqueducts bringing water into the city.

Despite the lead taken by the Romans, public access to sanitation and safe water did not become a priority for most countries until the nineteenth century. The usual motivation behind the removal of organic waste and sewage was the problem of odour, the desire for clean drinking water, and a dislike of wading through streets running with ordure. The direct connection of disease-carrying organisms with water pollution was not proven until the second half of the nineteenth century, when the germ theorists finally proved their case against the miasmists.

The towns and villages of medieval Europe seem not to have been very sweet-smelling places. Pigs were a convenient means of removing waste, and what they didn't eat would eventually be washed away by the rains. Many towns and cities had rudimentary regulations for the disposal of waste and teams of 'rakers' or 'scavengers' were often employed to remove garbage from the city. But by and large, the water management advances of the Roman Empire had been long forgotten.

Nevertheless, hard-pressed municipal administrations were already attempting to tackle water pollution problems in the early fourteenth century. An official investigation into the state of the Fleet River in London in 1307 concluded that the main cause of pollution was tanning waste and butchers' offal from Smithfield market.[11] In the same year, the Palace of Westminster installed a pipe connecting the King's lavatory with another sewage pipe that had been constructed earlier to remove waste from the palace kitchen.[12] Needless to say, this was not a privilege available to many commoners, and most people's sewage continued to flow direct from privies jutting over the river, or into the open gullies and trenches that ran down the streets. Sewers and cesspools were being developed, but their efficacy was doubtful. Sewers were often blocked (and in any case

4

simply emptied into the nearest river or stream), while cesspools stank, overflowed and tended to leak into neighbours' wells.

Fines for pollution were already being levied in London at this time. A 1306 proclamation on air pollution from coal threatened offenders with 'grievous ransoms'[13] and by 1345 householders could be fined two shillings for not removing filth and refuse from outside their dwelling.[14] In the late fifteenth century the secretary to the Venetian ambassador wrote with amazement of the existence of laws banning the killing of ravens and kites, because 'they keep the streets of the town free from all filth.'[15]

Pollution and the Plague

By most accounts, the medieval world was more conscious of sanitation than the later renaissance civilization, but it didn't prevent Europe succumbing to bubonic plague. Philip Ziegler chronicled the impacts of the great pestilence as it swept through Europe, in his classic book *The Black Death*.[16]

The plague of 1347 was a disease caused by a bacteria carried by fleas and spread by rats. It came in the wake of widespread death and starvation caused by soil exhaustion and population growth and compounded by extremes of cold weather and high rainfall. The rats themselves had probably arrived in the boats of crusaders returning from the middle east, and they flourished in the overcrowded and unsanitary conditions of medieval Europe. The famine-weakened populace and the virtually uncontrolled flow of sewage through the streets and into the rivers of most European cities provided ideal conditions for the Black Death. Unchecked in its spread, the pandemic probably killed a third of the people in Europe over a period of just two and a half years.

The medical knowledge of the time was unable to explain the plague, but the idea of 'corruption' of the air held sway among doctors. Ziegler credits Ibn Khatimah from Granada with the idea that 'the very nature of the air might be permanently changed by putrefaction'. Alfonso of Cordoba believed someone was waging germ warfare 'since air can be infected artificially', and that the evil doer needed only to prepare the dastardly 'confection' in a glass flask and then break the container into the wind so that 'the vapour pours out and is dispersed in the air'. Ziegler himself says 'almost every fourteenth century savant or doctor took it for granted that the corruption of the atmosphere was a prime cause of the Black Death'. Although completely wrong, like the second century Greek physician Galen before them, these medieval medics were laying the conceptual foundations for the study of invisible air pollution.

The plague also precipitated a pogrom against the Jews. Despite the fact that most people believed the disease to be spread through the air, the Jews were blamed for deliberately poisoning the wells. Through the winter of 1348 and on into the next year, the citizens of town after town in central Europe massacred Jews. In Basle they were penned in wooden buildings and burned alive; 600 Jews were killed in Brussels and similar inhumanities occurred in Stuttgart, Freiburg, Dresden, Erfurt, Barcelona and a host of other cities.

In trying to explain the blame that was so brutally conferred on the Jews, Philip Ziegler suggests:

> many wells were polluted by seepage from nearby sewage pits. The Jews, with their greater understanding of elementary hygiene, preferred to draw their drinking water from open streams.... Such a habit, barely noticed in normal times, would seem intensely suspicious in the event of plague.

Dealing with Waste

The problem of management of the disposal of human sewage and organic waste persisted well into the nineteenth century in most colonial nations and still curses much of the Third World. Clive Ponting describes how in 1366, the butchers of Paris were made to dispose of their animal wastes outside the city; how the same citizens, prior to the revolution, used a row of yew trees in the Tuileries as a *pissoir* and how the method of street cleaning in Madrid at the end of the seventeenth century was merely to empty barrels of water and 'let the filth run off.'[17] The water closet, or WC, was invented by an English poet, Sir John

Harrington, in 1589 but because of Elizabethan England's complete indifference to dirt and its lack of sewage piping, this sanitary innovation was ignored. It was not until 1778, when Joseph Bramah began marketing his own patented closet, that the use of WCs began to be taken seriously.

The unsupportable stench of city pollution was a commonplace throughout Europe for centuries. Travellers to Narbonne in France had dubbed the city 'Latrina mundi, cloaca Galliae' by the mid-seventeenth century. And in 1772, Pierre Patte described how in cities such as Bordeaux, Lyon and Toulouse, all manner of filth 'washed openly through the gutters before it reaches the sewers...then you have the blood from the slaughterhouses streaming through the streets.'[18]

A few decades earlier, in 1711, Jonathan Swift had published a lament on the disgusting mix of detritus that oozed from London's drains during rainstorms. He wrote of the 'sweepings from butchers' stalls, dung, guts and blood, drowned puppies, stinking sprats, all drenched in mud, dead cats and turnip tops' that swept through the city on the flood.[19]

Nearly two centuries later a remarkable account of the state of Hamburg's drinking water system (when the Elbe used to flow unfiltered through it at the end of the nineteenth century) was presented by Anna Hastings, 'of the water pipes of a great city encrusted almost throughout with sponges, polyzoa and molluscs, sheltering a dense population of crustacea and worms, and with many thousands of eels swimming in the fairway.'[20] Water filtration had, however, been installed by 1892 in the neighbouring Prussian town of Altona whose boundary with Hamburg lay along a city street. And when Europe's last great cholera outbreak struck in 1892, the disease swept through families on the Hamburg side of the street and completely spared those on the other who were drinking purified water.[21] This episode neatly demonstrated the link between clean water and health and prompted the Hamburg city fathers at last to treat the Elbe waters.

Not everywhere was in the same sorry state, however. The Chinese had long had a system of waste collection and the spreading of 'night-soil' was an essential part of the agricultural system that kept the fertile east China alluvial plains under productive agriculture for 4000 years. The tradition continues in many Chinese towns today, and Han Suyin wrote that 'in the city of Chengtu in the nineteenth century, and up to 1949, some of the wealthiest families were those owning the public cloacae and selling excrement to the countryside.'[22] A similar but less rewarding occupation could be found in eighteenth century London according to Robert Hughes, 'that of the "pure-finders"...old women who collected dog-turds which they sold to tanneries for a few pence a bucket.'[23]

Cholera, the driving force behind most European public health reforms in the nineteenth century, reached North America and Europe for the first time in the 1830s. It had previously been endemic to Bengal and caused epidemics mainly on the routes of Hindu pilgrims to the lower Ganges. British troops were the first to spread the disease wider from about 1816, first by land to Nepal and Afghanistan, and then on board ship to Ceylon, Indonesia, China and Japan. By 1831, cholera was established in Mecca and travelled further afield via Muslim pilgrims.[24] Nine years later the British Parliament's Select Committee on the Health of Towns reported on supposed links between poor sanitation, drainage, ventilation and adequate water supply, and the prevalence of diseases such as typhus, cholera and consumption. In 1848, the year of a cholera epidemic that left 62,000 English people dead, *The Times* recorded that the disease was 'the best of all sanitary reformers, it overlooks no mistake and pardons no oversight'.[25]

Disease-carrying sewage wasn't the only contamination problem in Victorian waterways. Referring to the rivers Aire and Calder, the 1867 report of the Royal Commission on River Pollution spoke of 'water poisoned, corrupted, and clogged, by refuse from mines, chemical works, dyeing, scouring, and fulling, worsted and woollen stuffs, skin cleansing and tanning, slaughter-house garbage, and the sewage of towns and houses'. Three decades later the Tawe, as it flowed through Swansea, was polluted by 'alkali works, copper works, collieries, sulphuric acid liquid, sulphate of iron from tin-plate works and by town sewerage, slag, cinders and small coal'.[26]

Indeed, South Wales was a veritable witches cauldron of

industrial pollution. In the early nineteenth century the English Vivian family and the Anglesey mine owner, Thomas Williams, were able to turn the 75 hectare lower Tawe Valley into the world's most powerful metallurgical centre.[27] At the peak of the region's prosperity there were probably 400 chimneys belching smoke in this tiny Glamorganshire valley. In nearby Llanelli the world's highest stack (320 feet) was erected in 1861 to carry away fumes from the tiny town's huge copper works.[28]

The 1880s saw the Welsh copper industry being replaced primarily by zinc, but also lead, nickel, arsenic and silver, to be followed in the early twentieth century by tin plate and steel. For more than a century, local rivers were sterilized and forests died. As early as the 1830s farmers began to report the death of cattle and by 1888 the average age of people in Swansea was only 24. Similar depressing statistics could be cited for the rest of the country, and the impact of industrial pollution was breaking out like environmental eczema on the English landscape. But despite the pervasiveness of the duel blights of poor sanitation and heavy industrial pollution, Britain's first Public Health Act was not passed until 1875. It was followed 61 years later by the 1936 Public Health Act, which still listed occupations including blood boiler, bone boiler, fat extractor and melter, glue-maker, gut-scraper, soap boiler, tallow melter and tripe boiler as 'offensive trades'.

Apart from the health benefits of these Acts of Parliament, there were considerable positive changes to be seen in the fauna of the Britain's waterways. The return of grilse, whitebait, flounders, eel and smelt to the Thames estuary was recorded between 1895 and 1901. And the naturalist Richard Fitter credited these species' reappearance to the 'efforts of the London County Council's Main Drainage Committee, which instead of allowing all the sewage of London to fall into the river at Barking and Crossness, separated the solid matter and carried it out to sea'.[29] Today's environmentalists, campaigning against the continued dumping of sewage sludge by Britain (alone among North Sea nations) might not celebrate the achievements of the Drainage Committee so fulsomely.

Something in the Air

Air pollution has been a political issue in Britain for almost 800 years. When Queen Eleanor of Provence visited Nottingham Castle in 1257 the fouled atmosphere, full of heavy coal smoke, forced her to move to Tutbury Castle.[30] Numerous attempts to control coal burning and punish offenders were made during the thirteenth and fourteenth centuries, but largely failed. Queen Elizabeth the First was herself 'greatly grieved and annoyed' by coal smoke in the Palace of Westminster; a complaint which led the local brewers to agree to burn wood instead.[31] Widespread damage to vegetation was reported in the first years of the seventeenth century, as was soiling of household leather furniture and wall hangings.

In 1659, John Evelyn wrote that London was enveloped in 'such a cloud of sea-coal, as if there be a resemblence of hell on earth'.[32] After 25 years the situation had not improved and Evelyn wrote in his diary for January 1684[33] that:

London by reason of the excessive coldness of the air, hindering the ascent of the smoke, was so filled with the fuliginous steam of the sea-coal, that hardly could one see across the street, and this filling the lungs with its gross particles exceedingly obstructed the breast, so as one would scarce breathe.

Later still, towards the end of the eighteenth century, the naturalist Gilbert White observed that his beloved Selborne experienced 'a blue mist which has somewhat the smell of coal smoke, and as it always comes to us with a NE wind, is supposed to come from London. It has a strong smell, and is supposed to occasion blights'.[34]

In the Manchester of the 1840s, J G Kohl described how the 'numberless lamps, burning in the streets, sent a dull, sickly, melancholy light through the thick yellow mist' and in another contemporary account Engels declared, 'the smoke nuisance drives everybody from the township of Manchester who can possibly find means of renting a house elsewhere'.[35] The opening paragraphs of Charles Dickens' novel *Bleak House* speak of the November 'smoke lowering down from the chimney-pots, making a soft black drizzle, with flakes of soot in it as big as full-grown snowflakes – gone into mourning, one might imagine, for the death of the sun'.[36]

Similar scenes were to unfold throughout the industrial world. The centre of the world's meat industry, Chicago's stock-yards and slaughterhouses, must have been a gruesome place. After visiting them in 1904 the novelist Upton Sinclair wrote of smoke from the Packingtown chimneys that:

came as if self-imperilled, driving all before it, a perpetual explosion. It was inexhaustible; one stared, waiting to see it stop, but still the great streams rolled out. They spread in vast clouds overhead, writhing, curling; then uniting in one giant river, they streamed away down the sky, stretching a black pall as far as the eye could reach.[37]

In 1945 Richard Fitter wrote 'The pollution of the air by smoke and other noxious vapours is due to preventable causes, entirely under human control.... There is no more direct biotic influence of man...and no more easily remediable one, than atmospheric pollution.'[38] By 1990, more than three decades after the Clean Air Act was passed, smoke had been largely cleaned up and London's killing winter fogs were a thing of the past, but environmental campaigner Chris Rose could still write 'The UK boasts a whole network of giant power stations with "tall-stacks". They were built to rid Britain of smog, but in doing so they created exports of acid rain.'[39] Acid rain still permeates the atmosphere of Britain, soaking into the soils, corroding the health of forests and slowly strangling aquatic life in many areas. This poison from the air can trace an unbroken and polluted lineage centuries into the past.

Chapter Two

CITIES AS THE SOURCE OF POLLUTION

I wander through each dirty street
Near where the dirty Thames does flow
And mark in every face I meet
Marks of weakness marks of woe
                                    William Blake

Spawning Cities

Today's city with its millions of inhabitants, its web of tarmac, carrying trucks and automobiles, its offices, apartments and shops is a truly modern phenomenon. Nearly half of humanity lives in an urban environment, and in the industrialized North, hardly anyone really lives outside a town. It is the city which gives us our most enduring images of pollution, whether they be the smoke-stacks of nineteenth century England or the smog of Los Angeles. From the time when towns and cities began to multiply and grow, pollution began to exert its stranglehold on modern society. Cities provided a negative force on the old adage that 'dilution is the solution to pollution' – they concentrate people, their activities and commerce, and their waste.

It is hard to imagine the difference in scale between our own cities and those of previous generations. Both the phenomenal rise in gross population and the inexorable exit from the land have contributed to this. Athens, at its most successful in 431 BC contained some 300,000 people, while citizens of Rome five centuries later numbered about 650,000.[40] Whole civilizations revolved around these tiny hubs. The great Buddhist city of Taxila in the foothills of the Himalayas spread over just a few hectares. The harbour at the Punic port of Carthage, from which the Phoenicians controlled most of the Mediterranean, fought their battles with the Greeks and Romans, and ran much of the world's sea trade, was hardly bigger than an Olympic-size swimming pool.

Northern Europe was a late developer in urban terms. In the thirteenth century, Paris, Venice and Milan had almost certainly all broken the 100,000 population barrier. On the other hand, Germany's biggest city, Cologne, had a fifteenth century population of around 20,000 people, and London had only reached 40,000.[41] By the end of the sixteenth century, London was becoming Europe's commercial power house and had grown to quarter of a million people, closely followed in size by Milan, Naples and Paris, while Lisbon, Rome, Palermo, Seville, Antwerp and Amsterdam all had over 100,000 citizens.[42]

All these cities had their share of urban problems. Those of food and water supply, of disease and poverty, of traffic congestion and limitations on housing and energy supply. But despite this, the urban problems of the sixteenth and seventeenth centuries were not so far removed from those of the countryside. Even the big towns were not isolated from their surrounding environment. Gardens and fields could be found widely in cities, markets for fresh produce were a necessary part of town life, and urban dwellers relied on firewood for their heat just as did those in the country.

The Industrial Revolution

It was the eighteenth century that brought changes in social and economic life, for the mass of the population which would make a qualitative leap forward towards the city would prove a primary source of pollution. Nowhere was the change more apparent than in England, where the Enlightenment, the Inclosure Acts, the steam engine, a host of technological advances and the new breed of entrepreneur would usher in the factory age, causing a rush of rural–urban migration and closing the door forever on the old-style town and country relationship.

This new age brought the curtain down on the scattered out-working system where families or other small groups had been paid to carry out a range of operations required for the manufacture of goods such as cloth. Communities of out-workers for the textile industry, in particular, had flourished around sources of water for power and washing, such as those in the Lake District, the Pennines and the Peak District. Typically, families would be involved, with mothers and daughters carding and spinning yarn and the menfolk weaving. Travelling agents would collect the finished cloth for wages, and the families would probably supplement their small income by keeping a few animals and growing vegetables.

Conditions for industrialization started to come together in the second half of the century. But soon the Inclosure Acts were to wreak havoc on the lives of many of the rural poor. The first lands to be enclosed were the open fields of the Midlands and parts of Yorkshire, mainly for pasture in mid-century, and then a little later, the wastes and commons of the more southerly counties for grain. Those that suffered most were families that had relied on the commons for grazing, firewood and even shelter, as well as the cottagers who supplemented home-based or farm wage labour with livestock on the commons. These people lost a vital part of their support mechanism, and also the tenuous but valuable connection to the land as stake-holders. They became ripe fodder for migration to the cities.[43]

Meanwhile, the 'improvement' of land by drainage and ploughing continued apace, with heath, down, fen and pasture converted to arable. Thus could a growing urban population be better provided with food. The digging of canals enabled the convenient transport of bulky raw materials including coal, copper and iron ore, whilst the creation of turnpikes markedly improved the quality of roads and radically reduced travel time for individuals and goods.

The flying shuttle and spinning jenny increased the productivity of out-workers, but Arkwright's water frame, which enabled both warp and weft to be of cotton, and therefore the production of pure cotton cloth, needed larger premises. Arkwright built his first mill in Nottingham in 1770, and was able to do the work of 2500 out-workers with a mere 50 labourers. Mills multiplied and English spun cotton production increased from under half a million pounds in 1765 to 16 million pounds twenty years later. The factory had been born, but most of the mills still used water power.[44]

The invention of the steam engine allowed capitalists like Arkwright to break away from the natural power supplies of the fast-running rivers, and concentrate factories in and around cities. There were not more than 500 of Watt's engines in operation by 1800, but those that there were laid the foundation for the new concentration of heavy industry in the steadily growing cities, for the rapid and massive growth of the urban proletariat, and for the steady decline in living standards of the working people for much of the next century.[45]

The Spread of Squalor

In 1773 Manchester had a population of only 27,000, but by 1801 this had nearly trebled to 75,000.[46] As Asa Briggs has said, 'cotton made Manchester', with its new class of wealthy industrialists and merchants,[47] but it also made Manchester the first of the filthy cities. Engels, who lived there and knew it well, wrote in 1845 'If any one wishes to see in how little space a human being can move, how little air – and such air – he can breathe, how little of civilization he may share and yet live, it is only necessary to travel hither.' Of one area occupied mainly by Irish immigrants Engels wrote: [48]

The cottages are old, dirty and of the smallest sort, the streets uneven, fallen into ruts and in part without drains or pavement; masses of refuse, offal and sickening filth lie among standing pools in all directions; the atmosphere is poisoned by the effluvia from these, and laden and darkened by the smoke of a dozen tall factory chimneys.

People who had been poor or dispossessed in the countryside were flocking to the new urban centres where jobs were flourishing, but where mechanization was also taking place. Soon mechanization took over, wages stagnated and the cities provided the permanent source of surplus labour that industry thrived on. Cities grew around the new concentrations of 'manufactorys' with their smoking chimneys and constant streams of waste spewing into the rivers and canals. The burgeoning industrial towns like Leeds, Bradford, Halifax, Preston, Oldham, Bolton, Rochdale, Swansea, Wigan and Huddersfield were starting to match in growth and activity the older harbour cities such as Bristol, Liverpool and London, which had long-existing pollution problems. One London satirist devoted a whole poem to the state of the Thames in 1859,[49] ending:

> River, river, reeking river!
> Doomed to drudgery foul and vile;
> Noisome, noxious fumes distilling,
> Fumes which streets and houses filling
>   Harpy like, defile.

That same year, work began on London's main drainage system, which was aimed at relieving the Thames by diverting sewage across the West Ham and Barking marshes into the River Lea.[50] Of the rivers Medlock, Irwell, Aire, Tame, Irk, Tawe and many others we have been left innumerable contemporary accounts of the appalling pollution.

Engels described the 'unendurable stench' from the 'blackish-green slime pools' that were the River Irk in summer.[51] By the time Hugh Miller wrote that the Irwell was 'considerably less a river than a flood of liquid manure'[52] in 1862, the development of the railways had made worse the havoc wrought by the steam engine. The world of the late twentieth century is reeling from the urban degradation brought about by the spread of the motor car, and pines for the good old days of rail, yet many of the original urban planning sins were committed in the name of Old Man Rail. The railway brought about the final liberation of heavy industries from their geographic ties to the mines. Huge quantities of coal flowed into the towns on railroad tracks that split communities, and into sprawling freight yards that gutted town centres. Lewis Mumford claimed that 'Every mistake in urban design that could be made was made by the new railroad engineers, for whom the movement of trains was more important than the human objects achieved by that movement.'[53] The coal itself, burning in every factory, foundry and household grate, caused the worst concentrations of air pollution the world had known.

Conditions for workers in the industrial city worsened rapidly during the early part of the nineteenth century. Great slums grew up and ill health festered in the communities of back-to-back housing. These foul buildings, thrown up across the industrializing parts of Britain, had no through ventilation or sanitation, were cruelly overcrowded and often inhabited even in the cellars. The ill health that urban housing engendered was finally officially acknowledged when army recruiters looking for cannon-fodder for the Crimean War discovered how bad was the health of the city worker.

Although the cities of Britain gained a head start in the industrial revolution, the rest of Europe, America and many of the colonies were not far behind. Cities like Essen, Cologne, Mulhouse, Toronto and Melbourne began to grow explosively and form recognizable nuclei of pollution on the map of global progress. Melbourne's river Yarra was as polluted as any urban European water course by the late nineteenth century, and suburban mortality rates rivalled or exceeded those of London. For its growth out of nothing in the pioneering world of post-transportation Australia, the city had been lauded as 'the Chicago of the South', but thanks to its huge open sewers and lack of water treatment, it was also known as 'marvellous Smelbourne'.[54]

Taming North America

The real Chicago was something else entirely. Initially a product of the boosters that were opening up the West, the city grew from the mud and prairie to produce many of the great business innovations that would help create 'The American Way'. In his remarkable history of Chicago, *Nature's Metropolis*, William Cronon traced the development of the windy city from its beginnings as a fur-trading post in the 1770s to its Great Exhibition in 1893. This was a century of change during which Chicago was the central element in the redefinition of American commerce through the development of the lumber, meat-packing and grain trades. The city carved a niche from the natural landscape on the

17

shores of Lake Michigan to become the vital link between the old East and the new West, between consumers and their desires.

Asa Briggs has classed Chicago as one of the three great 'shock cities', nestled between Manchester in the 1840s and Los Angeles in the 1930s. He recorded Richard Cobden as saying in 1871, that the visitor should 'see two things in the US if nothing else – see Niagara and Chicago'. Rudyard Kipling was no less impressed, but having seen the legendary town 'urgently desired never to see it again'.[55] For with the stunning commercial success of Chicago came all the ills of urban and industrial pollution.

William Cronon[56] quotes novelist Frank Norris from his 1903 novel *'The Pit'*, describing the 'smoke blackening the sky' and the 'tempest breath' of the steelworks that gave Chicago a special place in America. 'Here, of all her cities', wrote Norris:

> *throbbed the true life – the true power and spirit of America; gigantic, crude with the crudity of youth, disdaining rivalry; sane and healthy and vigorous; brutal in its ambition, arrogant in the new-found knowledge of its giant strength, prodigal in its wealth, infinite in its desires. In its capacity boundless, in its courage indomitable; subduing the wilderness in a single generation, defying calamity, and through the flame and debris of a commonwealth in ashes, rising suddenly renewed, formidable and Titanic.*

The power of Chicago came from its situation, from its magnetism for innovative businessmen, and from the growing American appetite for natural resources to convert to capital. Where the technical innovations did not originate in Chicago, they were quickly taken up, adapted and made the city's own. Chicago culled slaughterhouse techniques from Cincinatti, grain elevators from Buffalo and gave a home to Mr McCormick's Virginia reaping machine.

In 1833, when the native tribes reluctantly signed away their homelands to the incoming settlers, a rapid population expansion took place, with Chicago's population rising from a few hundred to almost 4000 by 1886. Lumber yards had already sprung into existence in the 1830s, and the town's position on the lake gave it easy access to the great northern woods of hemlock, maple, elm, basswood and most profitably of all, the giant white pine. Soon the Chicago lumber industry would be providing the wood required to settle the timber-scarce prairies. Millions of fence posts were needed to turn these rich grasslands into private farms. Timber for houses, shops and churches. And timber for the railroad ties upon which the burgeoning rail network could be built.

By mid-century Chicago had become the world's biggest lumber market and shipments rose from 220 million board feet in 1860 to over a billion in 1880. The city's wholesalers grew rich on the profits to be taken between the loggers and the consumers out West. Some of this prosperity was based on another Chicago innovation – Augustine Taylor's balloon frame house. Rapidly constructed using newly mass-produced nails, the frame house revolutionized the building industry. Houses were not only quick to construct, but also portable. These skeletal frames did away with the necessity for heavy beams and carpenters skilled in the creation of mortice and tenon joints.[57]

Although Chicago's lumber yards were providing the raw material for the settling of the West, it was meat sent from those same farms that gave Chicago its distinctive smell. When settlers and Eastern hunters had destroyed the bison population, bringing it down from 30–40 million to just a few thousand in a matter of years, the plains became grazing for livestock. Although cattle came eventually to dominate the Chicago meat-packing trade, it was pork that started the ball rolling. Pigs were a great secondary crop for corn farmers as they could be fattened on grain and were able to put on additional bulk from free foraging and recycling of waste and spoiled food. But they were not good travellers, being difficult to drive across country and losing weight fast. Cattle, on the other hand, could be driven all the way to the East coast. People expected to eat fresh beef, but were used to pork preserved by salting or smoking. For these reasons, the meat-packing industry began with the hog trade.

Chicago took up and developed the 'disassembly line' techniques that originated in Cincinatti, and with the huge demand for meat in field rations that came with the Civil War, took a grip on the trade. From 20,000 pigs a year around 1850,

Chicago was processing in excess of a million animals annually by the start of the 1870s. From 1858 the city's pork packers were also able to turn what had been a seasonal trade into a year-round one by storing winter ice in huge warehouses. It was the development of refrigerated railroad cars a decade later that enabled Chicago for the first time to move seriously into the beef-dressing business. But not only was the ice-car hugely consumptive of ice (thousands of pounds of it for a four day journey), it suffered from the problem of pollution.

Ice cut in Chicago released the most disgusting stench as it melted in the packing houses and railroad cars, for the river itself was full of the putrid remains of the once-living pigs and cattle. As one citizen of Morris, Illinois, wrote in the 1870s, 'What right has Chicago to pour its filth down into what was before a sweet and clean river, pollute its waters, and materially reduce the value of property on both sides of the river and canal, and bring sickness and death to the citizens.' Filthy ice, foul smells and public protest drove the search for alternatives to ice (mechanical refrigeration), and also for uses of by-products that had previously been dumped in the rivers and sewers. So the packing companies diversified into glues, gums, dyes, margarine, fertilizer, sausage casings, brushes, candles and many other products. Alas, pollution just seemed to get worse as the trades expanded.

With lumber Chicago had helped to colonize the West; with meat the city was altering America's, and ultimately the world's, eating habits; and with grain a new system of trading developed. In the world of GATT, NAFTA, Lloyds and Wall Street, it is easy to forget that we have not long lived with a system of commodities speculation. William Cronon roots much of the responsibility in the invention of the grain elevator. Variations on an initial model used by Joseph Dart in Buffalo, Chicago's elevators were huge warehouses with separate bins for different grades of grain. Grain came in on scoops or buckets attached to mechanical conveyer belts and went out into waiting boats or railroad wagons, through chutes at the bottom of the bins.

The elevator enabled the trade to be speeded up enormously. Weight became the measure of grain, and a whole system of grades was developed which allowed different farmers' shipments to be mixed in the elevators. The Chicago Board of Trade developed as the arbiter of standards, and these standards in turn brought the ability to trade in futures. The arrival of the telegraph in the city in 1848 enabled information about grain prices to travel across country more rapidly than ever before.[58]

This was an extraordinary time for the development of the city of Chicago. In just a couple of years, along with the first telegraph came the first canal, railroad, stockyard and grain elevator. This was a city that was built on the trade resulting from the settling of the West. Capital was created from the resources of the northern forests, the prairies and the High Plains. A connection was forged from the transport links initially of the canals and Great Lakes, and then of the railroads. And new methods of business and technology were eagerly adopted or invented. The list of innovations associated with Chicago could go on. In 1872, Montgomery Ward started the first mail order business, and the first skyscraper soared upwards as the Home Insurance Building in 1883. While the railways allowed Chicago to grow and prosper, it was eventually replaced as America's second city by the third of Asa Briggs' 'shock cities', Los Angeles.

The Birth of Smog

If Manchester defined the city of the industrial revolution, and Chicago was the pushy newcomer of the Victorian age, Los Angeles is the motor city, or 'Coketown under a chrome plating' as Lewis Mumford put it.[59] Los Angeles was the birthplace of smog. Not the black, smokey, winter smog of Dickens' London, but the yellow-brown, poisonous, year-round haze now familiar throughout the cities of the world. The photochemical smog first described from Los Angeles is formed by the action of sunlight on nitrogen oxides (NO_x) and hydrocarbons (or volatile organic compounds – VOCs). This smog reduces visibility, damages crops and plants, and causes respiratory problems in young and old alike. Ironically, health-conscious joggers and cyclists are high risk groups because the smog reaches deeper into their lungs when they exercise outdoors. The major sources of NO_x and VOCs are car exhausts, and Los Angeles was where urban planning first gave free range to the car.

In 1961 Lewis Mumford wrote that the mass-suburbanization of Los Angeles had created an 'undifferentiated mass of houses, walled off into sectors by many-laned expressways...space-eating with a vengeance.'[60] He was writing the year after regulations came into force on cleaner fuels for Los Angeles. Ever since then, Los Angeles has been struggling to reduce the smog problem. Major steps forward have been achieved, but in not one year since 1955 has the Southern California government managed to bring ozone levels down below the Federal health standard of 0.12 parts per million. There are still parts of the city that experience 60 to 80 'Stage 1 Alerts' a year, when residents are advised not to engage in vigorous exercise outside. Los Angeles has 11.3 million people and 9 million motor vehicles. It is the world's largest market for gasoline.[61]

Urban planners went beserk in the United States after the Second World War. Building on the success of the oil and motor companies in closing down much public transport, the planners proceeded to build a landscape that it will take decades to alter. *Lebensraum* for the American family became the private space of the house and yard in the suburbs; the denser housing in the inner cities became undesirable as shops closed up and moved out. Ethnic minorities and poor communities have become ghettoized in these once prosperous neighbourhoods. US town planning today still revolves around the car, and the proliferation of out-of-town and warehouse shopping only serves to exacerbate the situation. Consumers are required to own a car in order to shop, and in most US city suburbs the population density is so low and the public transport network so skeletal, that a car is usually needed even to get to the nearest bus or rail station.

Los Angeles is the model of a sprawling, car-dependent city with an air pollution problem. In the 1950s it gave a foretaste of what was to come for other Western cities. But as we head towards the year 2000, by when it is predicted that half the world's population will live in cities, the problems are no longer confined just to the West. In the last four decades, while the population of Los Angeles has increased by 150 percent, Mexico City and Sao Paulo have grown more than sixfold (each to around 20 million people) and Seoul went from one million people in 1950 to ten times that in 1990.[62] Asian cities including Seoul, Jakarta, Bangkok, Guizhou, Manila and Beijing are experiencing phenomenal growth rates, and economic wealth is increasing buying power so that the beginnings of a motor vehicle ownership explosion in Asia is already under way. Photochemical smog is already commonplace in these cities, and in many developing country urban centres, pollution is worsened not by suburban expansion, but by the growth of slums at the urban margins. As OECD governments at last begin to get a grip on the pollution problems of their own cities, there is a new generation of urban pollution sources coming on-stream in an unprecedented way in the rapidly industrializing developing nations.

Chapter Three

WHY CARE? POLLUTION, NATURE AND ETHICS

Clouds of stored summer rains
Thou shalt taste, before the stains
From the mountain soils they take,
And too unlucent for thee make.

John Keats

Should the Polluter Pay?

Caring about pollution in relation to the environment is a relatively modern phenomenon. Today's environmentalists are largely rooted in the Romantic tradition: they see beauty in nature, a wilderness worth preserving, or species, each with an inherent right to exist untouched by man or at least protected from his worst excesses. Pollution seems qualitatively different to other problems. Deforestation, dam-building, mining, urban expansion, however, are all physical threats to the unprotected natural world, and can be kept at bay by the creation of protected areas and the erection of fences. But pollution seeps under barbed wire and falls from the sky beyond the limits of the highest brick walls. It flows down rivers and it can be found in the deepest recesses of the oceans or in the snows of the uninhabited Antarctic wastes.

Epithets such as insidious, invisible, creeping and miasmal, give descriptions of pollution an almost anthropomorphic edge. In the twentieth century, pollution is seen as an embodiment of humankind's struggle against nature. Forests smitten by acid rain, coral reefs choked by sewage and wildlife decimated by pesticides. Western advocates of environmentalism have a Velcro-like attachment to the idea of pollution as a crime against nature. Concern about human health, while widespread in the populace, is only marginal to the concerns of many environmentalists. When Washington DC slapped a tap water ban on a million people in the district (for fear of *cryptosporidum* contamination) for five days in December 1993, there was hardly a peep out of the environmental groups. Yet Washington probably has the greatest density of anti-pollution activists of any city in the world. It's just that they were busy stopping global warming and marine pollution.

But pollution in the past has been a catalyst for major social change. Even before the word itself was in common usage, pollution was a source of complaint. Before there were anti-pollution campaigns there were complaints about smoke, smells and every sort of unsavoury menace. But it was not really before the nineteenth century and the age of social reformers that pollution issues were transformed into issues of the public good on a widescale basis. And even to achieve this, understanding of disease and public health had to have evolved.

The debate about pollution revolves around definitions, ethics and attitudes towards nature. What is pollution? Is it wrong to pollute? Who is responsible? Who or what is suffering and how? Where pollution is obvious and affects few individuals the answers to these questions can be reasonably simple. So that even in the fifteenth century, if a tannery's waste was leaking into a neighbour's water supply, a local tribunal could easily rule on the culprit and some compensation. Today, the complex nature of the pollutants, the web of environmental legislation and the need for proof of harm might tangle the same two litigants for weeks, months or even years in court. Ethics too are at issue. A business that pollutes in pursuit of profit for its owners or shareholders, may believe that what it is doing is right. Local farmers whose land and livestock suffer from emissions will hold a different set of beliefs.

Away from issues of local pollution, much argument centres on the use and abuse of the global commons. These are the natural resources that everyone uses and should benefit from –

the oceans, the soil and the air. These resources are regarded as free goods by many. The so-called 'tragedy of the commons' is that while a few may benefit from the abuse of these resources, everyone must pay the price as no one individual or even nation bears overall responsibility. So as the oceans become more and more polluted, the industries that discharge their effluent, and the cities that pump sewage into the waters are acting as 'free-riders', sharing the costs of pollution with all those that share the seas. Problems like these have led to the development of international treaties and proposals such as the 'Polluter Pays Principle'. This latter concept suggests that those who cause pollution should be responsible for its clean-up, but is notoriously hard to implement.

Polluters often argue that it cannot be proven that the pollution in question originated with them, or that it caused the damage blamed upon it. If both these conditions can be satisfied, then arguments can continue for years about the financial implications of compensation. The mercury pollution in Minimata Bay, the Bhopal disaster and the wreck of the *Exxon Valdez* provide examples of where arguments over financial liability have dragged on for years. In Britain, the Central Electricity Generating Board and the Forestry Commission argued throughout the 1980s that acid rain from power stations was not the cause of forest die-back or lake acidification. The nuclear industry continually fights claims of radiation-related cancer; the pulp and paper industry employs scientists to argue that bleaching doesn't release dioxins to the environment; the pesticides industry spends millions to wage public relations war against those who say pesticides raise breast cancer rates and hormonally related birth defects. The business world has turned the burden of proof into an albatross around the necks of their opponents. The little guy, the co-user of the environment, hardly has a chance to win his arguments.

The primacy of economic growth, and the dominance of Western empirical science are partly to blame for this state of affairs. If business is forced onto the defensive over pollution, they have only to start threatening the loss of jobs and opposition starts to fall away. The 'polluter pays' often means that the consumer pays and people tend to care more about their expenditure budget than they do about pollution. Hence the failure of President Clinton's proposal for a tax on motor fuels in the US. Drivers did not want to pay even a few cents more on a gallon of petrol. In the world context this makes no sense as US per capita carbon emissions are higher than those of any other country, and their petrol prices far lower than anywhere else in the OECD. However, to an individual who has a five year loan on a new gas-guzzler stretching his income to the limit, the logic of fighting the tax is incontrovertible. And anyway, the oil companies have a handful of paid scientific consultants disputing, on TV and in the press, every new scientific finding on climate change, so Mr and Mrs America are not even convinced there is a problem. After all, what proof is there that climate change is going to happen?

Science, Environment and Romanticism

The search for empirical proof that an environmental problem exists is rooted in the scientific developments of the sixteenth and seventeenth centuries. On the other hand, the modern environment movement depends for its support on the Western romantic notion that nature is beautiful, spiritual and inherently worth preserving. Nature has a special value and contemplation or direct experience of it fulfils an inner human need. As Wordsworth said:[63]

> One impulse from the vernal wood
> Will tell you more of man
> Of moral evil and of good
> Than all the sages can.

For most of human history, nature has been something to be used or mastered, often feared and sometimes worshipped. Growing populations in Europe, along with scientific and philosophical advances leading up to the eighteenth century, brought about major changes in human relationships to the countryside. As the western world emerged from the Middle Ages, cities grew and trade strengthened among different cultures. Desire for new commodities like sugar, alcohol, pepper, glass and linen was springing up. There were more mouths to feed and meat consumption actually went down after the middle of the sixteenth century, but people wanted better bread and grain prices rose. Hunger was commonplace throughout Europe until the eighteenth century, and the majority of people were extremely poor and almost without possessions. According to Fernand Braudel 'the poor in the towns and countryside lived in a state of almost complete deprivation.'[64]

The slow climb from almost universal poverty which allowed the spread of household items such as chairs, tables, hearths and stoves, forks and even pottery, as well as a greater variety in diet, required the transformation of the landscape. In a world where food was scarce and belongings meagre, reflection on the beauty of nature was almost non-existent. Winter cold was a constant foe in Europe and the climate an arbiter of well-being in its influence on agriculture. Untouched nature, in the form of marshes, woodlands, mountains or wild rivers presented a series of obstacles to travellers, traders and farmers. The long and ancient ridgeway tracks of England, for instance, had been used for centuries to avoid woods and marshy valley bottoms. Fear of wild animals or bandits and the sudden calamities that could be brought by flood, drought or avalanche were widespread. As Grevel Lindop has succinctly stated, 'Nature is beautiful as long as we are safe within it.'[65]

Lindop identified three stages in the Romantic response to nature and pinned them down as being exemplified in the Lake District writings of Thomas Gray, Coleridge and Keats between 1769 and 1818. 'In this brief period', he writes, 'our culture's way of looking at the non-human world turned a corner.' Gray's view of Derwentwater as the 'sweetest scene I can yet discover in point of pastoral beauty' encapsulated the emerging idea that nature could be beautiful if viewed as a scene or picture. Thirty years later, Coleridge enjoyed a more moving experience looking at Bassenthwaite Water from his house. Says Lindop 'For Coleridge, the landscape (though still framed by the windows) has entered the mind: it is now an experience, a state of perception, a tranquillizing or intoxicating dream tasting of the creative imagination itself.' Finally Keats' assertion that the landscape's 'countenance or intellectual tone must surpass every imagination and defy every remembrance' speaks to Lindop of an implied intelligence in nature – 'The implication is there that we must learn from this; that it knows, in some sense, more than we do.'[66]

To the Romantics, a human being could but learn from and be spiritually reawakened by contact with nature. For example, extolling the splendour of dusk at Rydal Mount, Wordsworth wrote:[67]

> *My soul though yet confined to earth*
> *Rejoices in a second birth.*

Similar echoes can be heard in the voice of the so-called 'Peasant Poet', John Clare:[68]

> *Nature thou truth of heaven if heaven be true*
> *Falsehood may tell her ever changeing lie*
> *But nature's truth looks green in every view*
> *And love in every landscape glads the eye.*

The Romantic tide in thinking and the arts swept through Europe and America, and the ripples spread in many directions. There was not one stream of Romantic thought, but many. 'Nature is Imagination itself' said William Blake, and with others such as Rousseau and Alexander von Humboldt raised the appreciation of the wilder countryside to an almost religious level. However, as Keith Thomas has written, it was the English that excelled in the 'divinisation of nature'. They streamed into the newly accessible wildernesses of Snowdonia and Cumberland, and they formed the bulk of the new breed of alpinists that flocked to seek spiritual enlightenment while climbing and botanizing among the previously abhorred peaks of Switzerland, France, Italy and Spain.[69]

In America, the transcendental movement gained momentum. Centred upon the thoughts and writings of Ralph Waldo Emerson and Henry David Thoreau, they sought inspiration in the organization and force of nature. This veneration of nature was supported by the writings of Georges Buffon who used geology to suggest that fixed biblical time was false, and that the Earth might be 75,000 years old at least. It built, too, on the philosophical works of Immanuel Kant and Johann Gottfried Herder which for the first time offered unified cosmic histories that accepted evolution over unlimited time and a complex web of relationships between nature and society.[70]

In Thoreau's writings there is a profound sense of trust in nature as the means to balance man's empty materialistic urges. In *Walden* he writes, 'I went to the woods because I wished to live deliberately, to front only the essential facts of life, and see if I could not learn what it had to teach, and not, when I came to die, discover that I had not lived.'[71] Human solitude, he felt, could be turned to positive result in lifting the spirit in communion with nature – 'Yet I experienced sometimes that the most sweet and tender, the most encouraging society may be found in any natural object, even for the poor misanthrope and most melancholy man. There can be no very black melancholy to him who lives in the midst of nature and has his senses still.'[72] Emerson too, in *The American Scholar*, called for the recognition of the vital role of nature for man's being. 'The first in time and the first in importance of the influences on the mind is that of Nature.'[73]

The romantic revolution in cultural and spiritual thought was not taking place in a vacuum. A huge variety of outside influences was driving people towards a reassessment of their relationships with the natural world. Jacob Bronowski reminds us that Blake 'lived in the most violent age of English history'. He lived contemporaneously with the Seven Years War, the American War of Independence, the French Revolution, the Napoleonic Wars and the Peterloo Massacre. During Blake's lifetime, the population of England doubled, as did the cost of living, but wages stayed the same and workers' rights eroded fast with the passing of the Corn Laws, the Inclosure Acts and the Poor Laws.[74]

Reactions to Reality

Great changes were taking place in society and in the landscape, and science was transforming the physical and intellectual background against which the game of politics was being played. As the industrial revolution gained strength, industry moved into the factories and towns. Agriculture was becoming more mechanized and millions of acres 'improved' and enclosed. The building of roads, the development of canal systems and then the coming of the railways shattered previous notions of mobility and speed.

Sociologist Colin Campbell believes Romanticism can be seen partly as a development from, and reaction against, the Enlightenment and what John Stuart Mill described as 'the narrowness of the eighteenth century'.[75] This narrowness was one of the results of the scientific revolution begun by Copernicus in 1543 when he suggested that the earth revolved around the sun and not vice versa.

The revelation stood medieval cosmography on its head and opened the way for the development of the science we know today. Keplar and Galileo developed the mechanistic world view which held that nature was governed according to mathematical principles. The logical extension was to empiricism, where everything real could be measured. Descartes contributed the reductionist theory that the universe was full of matter that was infinitely divisible into particles and that these particles could be described in mathematical terms. Plants and animals, even humans, were basically machines defined essentially by the complexity of make-up and motion of their particles. In separating man from nature, by giving the mind secondary qualities that were not objectively knowable, Descartes made him a rational being and thereby, different.[76]

Francis Bacon was the father of the scientific tradition that still holds sway in society today. His science was based upon the collection of facts and data, or empirical knowledge. With this knowledge, analysis would be possible, the laws of nature deduced and ultimately power could be asserted over nature. Newton's discoveries at the end of the seventeenth century were the key to unlocking this power and he described principles that were universal in nature.[77] This was the rational science which grew to dominate western thinking in the eighteenth century, and it was to its basis and its results that the Romantics were reacting. Man and his endeavours had been reduced to the level of a machine and its output.

The Romantics were seeking something more from nature. They were not in tune with Descartes or Newton and wanted to put the mystery back into their surroundings, to believe that

31

there was an inherent unknowability about the complexity and beauty of nature. As Peter Marshall has suggested, William Blake was 'an ecological poet *par excellence*', and other Romantics such as Shelley, Coleridge and Wordsworth were highly sympathetic to his exhortation[78]

> *To cast off the rotten rags of Memory*
> *by Inspiration,*
> *To cast off Bacon, Locke, and Newton*
> *from Albion's covering,*
> *To take off his filthy garments and clothe him*
> *with imagination.*

Faith in imagination was a thread running through the Romantic movement. Again, Blake encapsulated the links when he said 'Some see nature all ridicule and deformity...and some scarce see nature at all. But to the eyes of the man of imagination, nature is imagination itself. As a man is, so he sees.'[79] According to South African writer Breyten Breytenbach:[80]

> *this means that nature and man cannot be dissociated, and that we are as responsible towards nature as we would be to the scope and the sense of human existence. A person without imagination must go mad. A people without imagination will by definition be destructive.*

This destructiveness was apparent to the Romantics as they went about the business of discovering nature. Their world was changing fast and they were not affected less than any persons of their time. To a certain extent they were driven to nature as an escape from the increasingly populated and polluted towns. Wordsworth campaigned against the coming of the railway to lakeland because the quiet villages and hills would become overcrowded and vulgar. Of the wider transformation of the countryside he wrote:[81]

> *I grieve, when on the darker side*
> *Of this great change I look; and there behold*
> *Such outrage done to nature as compels*
> *The indignant power to justify herself;*
> *Yea, to avenge her violated rights.*

That changes had been occurring for many decades was a matter of record. For instance, in 1776 Gilbert White wrote of the disruption of watersheds, 'since the woods and forests have been grubbed and cleared, all bodies of water are much diminished; so that some streams, that were very considerable a century ago, will not now drive a common mill.' A few years later he noted that the conversion of beech woods to agricultural land had severly reduced populations of wood pigeons in the Selborne area.[82]

John Clare, himself the son of a Northamptonshire thresher, lived through the enclosures of common land, and wrote movingly of the impacts the political changes were having both on the rural poor, and on the land itself. In one poem, 'Helpston',[83] describing the lost rural idyll of his birthplace he wrote:

> But now alas those scenes exist no more...
> Now all laid waste by desolations hand
> Whose cursed weapon levels half the land
> Oh who could see my dear green willows fall
> What feeling heart but dropt a tear for all.

The Inclosure Act for the parishes of Helpston and nearby Maxton was passed by Parliament in 1809 and with it came not only hedges and forest clearance but also extensive drainage of the local wetlands. Bogs, marshes and water meadows had been regarded for centuries as dangerous, poisonous lands that could be cleaned up and turned to productive agriculture through drainage. During his travels in the Fens at the end of the seventeenth century, Daniel Defoe spoke enthusiastically of drainage, and expressed himself 'longing to be delivered from fogs and stagnate air, and the water the colour of brewed ale.'[84] Even the wetlands of the English midlands harboured malaria until the late nineteenth century, and marsh miasma was thought to be the cause of the ague, literally, 'bad air'. The destruction of the countryside and the channelization of the local river were recorded by Clare[85] in 'Remembrances':

> Inclosure like a Buonaparte let not a thing remain
> It levelled every bush and tree and levelled every hill
> And hung the moles for traitors – though the brook is

running still
It runs a naked stream and chill.

Some years later, Thoreau too, in quite different conditions on the far side of the Atlantic, was aware that his world was rapidly taking another shape, and that a retreat to the woods was an act of temporary respite and self-renewal. The Fitchburg Railroad track edged onto Walden Pond and Thoreau's woods were rent with the screams of locomotive whistles as the new commodity trades were developed and the cities sucked the country dry of its natural resources. 'All the Indian huckleberry hills are stripped', he wrote, 'all the cranberry meadows are raked into the city.' The smoke from the engines daily blotted out the sun and cast the fields of Concord, Massachusetts under shadow.[86]

> *Men think that it is essential that the nation have commerce, and export ice, and talk through a telegraph, and ride thirty miles an hour.... If we do not get out sleepers, and forge rails, and devote days and nights to the work, but go tinkering upon our lives to improve them, who will build the railroads? And if railroads are not built, how shall we get to heaven in season.... We do not ride on the railroad; it rides on us.*

Thoreau, Blake, Wordsworth and Clare, from their very different perspectives were chronicling the decline of the importance of the individual in society while simultaneously trying to reverse the trend. Thoreau believed, 'To affect the quality of the day, that is the highest of the arts' and a century later Bertrand Russell brought a modern echo to this sentiment when he stated 'A good society is a means to a good life for those who compose it, not something having a separate kind of excellence on its own account.'[87]

Russell also recognized that rationalism and Newtonian science was behind both the good and the bad of the modern age. He saw that scientists had fought authority as they sought to overturn established beliefs in past centuries, but that by the mid-twentieth century, they were in the hands of politicians who would seek to make use of the 'terrifying power to control nature.'[88] He rightly predicted that a modern anti-scientific movement might be spawned by the fear of nuclear and germ warfare.

In reality, that anti-science movement had begun with the Romantics, grew through all the scientific and engineering glory of the industrial revolution, and has its current home in the strand of the environmental movement that is represented by the 'deep ecology' made popular by writers such as Arne Naess and Edward Abbey. The publication of Abbey's *Desert Solitaire* in 1968 caught the growing mood of disenchantment with modern urbanization and technological development. Abbey revered the outdoors. He wrote of wilderness that 'It means something lost and something still present, something remote and at the same time intimate, something buried in our blood and nerves, something beyond us and without limit.'[89] Deep ecology rejects humanism and embraces a philosophy based on the intrinsic right to life of all species, and is ecocentric in asserting their equality to man. Pollution in this thought system is seen as a crime against other species and against nature.

The majority of the environmental movement today has been drawn into the camp of Bacon and Newton, while struggling to retain a Romantic connection with nature. Utilitarian arguments backed with empirical data are far more likely to win the day in any anti-pollution battle than is an appeal for a holistic approach that protects nature for its own sake. The Copernican revolution not only put an abrupt end to astrology and alchemy around three hundred years ago, but it also prepared the world to resist deep ecology. Even environmentalists who use science as a means to achieve their ends of conservation are often dubbed 'emotional', 'weird' or 'ignorant'. Always they are portrayed as the enemies of progress. Rationalism does not allow for the unknown or the unproven. It militates against cleaning up the pollution of the seas because we cannot prove a link between tumours on fish and a particular chemical. It provides a justification for nuclear bomb tests because the ability to wage a war with missiles is empirically proven but the epidemiology of radiation-related sicknesses is not. Canadian author John Raulston Saul has said 'Doubt is seen as a sign of ignorance. It is the enemy of the solution addicted elites.'[90]

Chapter Four

THE FIRST CONSUMER REVOLUTION

*And how much for
the brain of a pocket mouse,
and how much for
the hump of a humpback whale?*

Miroslav Holub

Climbing the Ladder of Consumption

The vastly inequitable consumption of resources by people in the industrialized nations is the basis upon which our environmental crisis is built today. The burning desire to possess and consume was only truly unleashed during the last 50 years, but centuries of attempting to slake human thirst for material and non-material goods have driven the planet into a downward spiral of resource depletion and pollution. The wealth of the western world has allowed its inhabitants to buy hundreds of millions of durables such as televisions, videos, cars and books in the last few decades, but what was the role of the consumer in earlier times?

It is a mistake to believe, as many do, that the roots of the resource consumption and pollution problem are only planted in the industrial revolution of the late eighteenth century. From the myriad historical points when agricultural produce began to be traded and craftsmen began to produce goods for barter or sale, came the problem. Its seeds lie deeper still, in human nature itself, because as Aristotle said, 'the avarice of mankind is insatiable'. And Adam Smith believed 'every man is rich or poor according to the degree in which he can afford to enjoy the necessaries, conveniences, and amusements of human life'.[91]

Although certainly not the first example of a period of increased consumption among a human population, the colonial expansion of Europe from the late fifteenth century is a good launching point for this account. Columbus's 1492 landfall in the Caribbean marked the beginning of the destruction of the native peoples of the Americas. The subsequent bloody expeditions of generations of European invaders were inextricably linked with the desire for wealth and possession of exotic wares. Columbus himself had set off to the West across the 'Sea of Gloom' in the hope of reaching the Indies and reopening the flow of consumer goods such as silks and jewels and particularly spices, which had been plugged by the Turkish victory in Constantinople in 1453.[92]

Historian John Hemming recorded Bartolome Ruiz's first contact with the Inca civilization, when as leading a forward party for Francisco Pizarro's 1526 expedition, he captured a balsa trading raft. The report sent back to King Charles I said of the native's boat 'they were carrying many pieces of silver and gold as personal ornaments...tweezers and rattles and strings and clusters of beads and rubies...pieces of clothing coloured with cochineal, crimson, blue, yellow and all other colours.... They were taking all this to trade for fish and shells....'[93]

By the middle of the sixteenth century, Peru had developed two great trades: the growing of coca leaves for the internal market, and the production of silver. The conditions in the silver mines of Potosi were appalling, and the ore that was extracted to finance the expansion of the Spanish empire was dug at the cost of many lives. Until 1571, the silver ore had to be smelted in wind furnaces, but with the discovery of mercury and the perfection of a process for refining silver using this toxic metal, new horizons were opened up to the Spanish masters of Peru.

Mining was an ancient occupation, and the Romans suffered sorely from the side effects of extracting silver from galena, or lead ore. Pliny the Elder recognized the potentially harmful effects of lead on the Roman population, and recent analysis has borne this out. During the heyday of the lead smelting industry, for about 400 years from 200 BC, Roman bone lead levels were consistently ten times higher than those of today. Apart from the direct impacts of local run-off and pollution from the mines and smelters, lead was in common use for water pipes, plates, drinking and storage vessels. The fall of the Roman Empire has been attributed by some to lead pollution.[94]

In 1550 the German metallurgist Georgius Agricola had written of the destructive impacts of mining and how the practice of ore washing 'poisons the brooks and streams and either destroys the fish or drives them away'.[95] For sixteenth century mineworkers too, theirs was a forlornly hazardous occupation and the conditions in the Huancavelica mercury mine were a vision of hell. There were no safety precautions, nothing but candles for light and a complete lack of ventilation in the deeper parts of the mine. Describing the pollution, Hemmings writes:

> When the Indians broke the hard, dry ore with their crowbars they were struck by a thick, toxic dust that contained no fewer than four poisons: cinnabar (sulphide of mercury), arsenic, arsenic anhydride and mercury vapours. These caused severe damage to the throats and lungs of Indians already debilitated by heat, exhaustion and bad diet.

Eventually, many of them died coughing up blood and mercury.[96]

These kinds of exploitative activities were the basis of much of the wealth extraction processes of the European colonial era. The development of international trade on a global scale, whether the commodities were slaves, sugar, rubber or any one of a host of others, allowed the flowering of a middle class in Europe, a class of consumers. The introduction of many new consumables, most of them natural products like foodstuffs, spices or plant dyes, into the markets of Europe from about the year 1600 onwards helped to develop a culture of experimentation with new products and, perhaps more importantly, a buying habit.

Tobacco was one of the newly available consumer items at the end of the sixteenth century. First described by Europeans returning from the New World in 1497, it was introduced initially to the court of Philip II of Spain in 1559 and then by Jean Nicot to the French court. Sir Walter Raleigh is credited with promoting its popularity in England.

Ironically, tobacco smoking was regarded in these early days as a cure for disease. More than four centuries later Kenneth Mellanby wrote 'most other types of air pollution pale into insignificance before the one really serious form – cigarette smoke'.[97] Despite efforts by the church and other groups to put an end to tobacco smoking towards the close of the seventeenth century, the buying habit had well and truly stuck. In Europe today, per capita consumption of cigarettes is over 1600 per year, and in many developing countries, including Thailand, Indonesia and Columbia, spending on tobacco has more than doubled in the last decade.[98]

Creating the Demand

What else could the consumer be persuaded to buy? As the gold and silver ran out, there was an increasing need for the countries of Europe to maintain the flow of capital into their exchequers through a positive foreign trade balance. Under foreign administrations, the role of the colonies was, virtually without exception, to provide a cheap source of raw materials such as timber, furs or ores, which could be converted in the home country to goods with high value added and re-exported at a profit, to foreign markets.

In 1776, two things happened which would have major consequences for the future. Firstly, Adam Smith published the *Wealth of Nations*, in which he supported the free market, minimal government control, and the idea that the pursuit of self-interest would have positive impacts on the general welfare. And secondly, America freed itself from the yoke of English rule. This latter event allowed the flowering of the American entrepreneur, in much the way Adam Smith had predicted. In doing so, the spadework was done for our own consumer revolution of the second half of the twentieth century. Within a few years of the declaration of independence, American traders, mostly from the seafaring state of New England, were swarming the globe. As Historian Daniel Boorstin put it, 'the greatest resource of New England was resourcefulness'. He went on to recount the exploits of an American entrepreneur whose efforts link directly to the development of modern marketing and, through the icebox, to the hole in the ozone layer.

In 1805, Frederic Tudor from Boston began to build a business selling winter ice from New England ponds. In developing a consumer demand for ice to cool drinks, preserve food and make ice cream with, he was creating a market out of thin air. He built ice houses in Charleston and Havana, and in the 1830s in Calcutta, says Boorstin, he built an 'ice depot and encouraged Anglo-Indians to buy household refrigerators and water coolers; he tried to change their eating habits by his well-preserved shipments of apples, butter and cheese.'[99]

Chemical refrigeration had first been tried in Rome in the 1540s, water-ice (or sorbet) was developed in Florence around the same time, and dairy ice cream occasionally turned up on well-heeled European tables from the middle of the seventeenth century.[100] It was in the warmer climate of America, however, that the demand for coolboxes and ice really grew. Ice consumption in cities including New Orleans, New York and Boston rocketed in the early decades of the nineteenth century.

Boston became the centre of the world ice trade, with fortunes being made from the marketing of a natural and renewable commodity. Tudor had precipitated 'the Ice Age of American diet – with its emphasis on sanitation, nutrition and refreshment' says Boorstin. Today, New England ponds such as Henry David Thoreau's beloved Walden and White Pond, of which he was able to say, 'they contain no muck',[101] are polluted with chemicals best not served in an iced drink. But the revolution Tudor started through the promotion of ice has ramifications today. Without the popularization of ice there would have been no Budweiser, no Baskin Robbins and no Burger King. Clarence Birdseye could not have developed a market for his frozen food, Eugene O'Neal would not have written *The Iceman Cometh* and neither James Bond nor Hawkeye Pearce could have indulged in Martinis.

The icebox was followed by refrigeration and air conditioning and the fridge by the freezer. The ability to cool, chill or freeze any product or space is now regarded as a basic need in modern society. Unfortunately, the chlorofluorocarbons (CFCs) used since the 1950s for both domestic and commercial refrigeration have been the main cause of the destruction of the earth's protective stratospheric ozone layer.

As J K Galbraith has observed, 'When man has satisfied his physical needs, then psychologically grounded desires take over. These can never be satisfied or, in any case, no progress can be proved.'[102] So we must look again to the development of the new consumers, who with a disposable income and the desire to be different from their neighbours and peers, formed the backbone supporting the new production, and hence, the rise of industrial pollution.

But what were the items that tempted the new consumer? Jordon Goodman and Katrina Honeyman identify the key products as 'clocks, watches, bicycles, sewing machines, typewriters and small arms.'[103] To this list could be added ready-made clothing, cameras, convenience food, record players, telephones and lightbulbs. Asa Briggs has highlighted the novelty of the now commonplace in the Victorian world. Mass-produced steel pins and needles, fountain pens, corrugated iron, crinoline, aniline dyes and vulcanized rubber were just some of the host of new products to excite and eventually captivate the consumer.[104]

During the last forty years of the nineteenth century, an extraordinary consumer boom took place in the northern hemisphere, and America replaced Europe as the centre of innovation and invention. Fresh off the boats and freed from the suffocating embrace of history, religion and property, European immigrants truly were setting foot in a land of opportunity where hard work, good luck and talent could be converted into American prosperity.

Grinding poverty for the immigrants was the norm, but wages could be exchanged for socially equalizing clothing and myriad new consumer products that provided material proof of escape from the class-ridden hierarchy of dress codes and ownership rules in Europe. Shopping became a search for freedom and self. As Jonathan Raban[105] has commented, 'The windows of the department stores were theatres. They showed American lives as yet unlived in, with vacant posession. When

41

your nose was pressed hard against the glass, it was almost yours, this other life that lay in wait for you with its silverware and brocade.'

Immigrants and their sons and daughters, with names like Berliner, Bogardus, Bonner, Singer, Pulitzer, Prang, Swift, Gair and Gantt were at the centre of the inventive frenzy of post-civil war America. Made possible by the development of mass production techniques and spurred on by new marketing methods, this era saw the conception of most of the basic conveniences of everyday life that people in industrialized countries take for granted even now. In the packaging field, for example, canning of food and the production of condensed milk had become commonplace by the 1860s, paving the way (with refrigeration) for long-distance movement and long-term storage of foodstuffs. During the same decade, American paper bags revolutionized shopping and by 1879 the machine-made folding boxes that were the precursors of packages for everything from cigarettes to cornflakes were invented. Collapsible metal packaging was being developed at the same time, and toothpaste was being marketed in disposable tubes by 1892.

Meanwhile the consumer was being tempted to buy the new products in ever more ingenious ways. Emile Zola famously commented on the way in which department stores 'democratized luxury', and the Paris store Bon Marché took the haggling out of buying, with a fixed price policy in 1852. The Great Atlantic & Pacific Tea Company became the first chain store during the 1860s. Plate glass opened the way to window shopping, F W Woolworth opened his first 'Five and Ten Cent Store' in Utica, New York in 1879. The 1880s saw Montgomery Ward and Richard Sears and Alvah Roebuck pioneering the mail order catalogue with customers throughout rural America.

The development of the new specialisms of advertising and market research as well as the marketing of consumer festivals such as Christmas, all strengthened the drive towards consumerism. Thomas Nast popularized the modern image of a white-bearded Santa Claus in drawings for *Harper's Weekly* in 1863. In 1891 Mr Woolworth was able to tell his store managers before Christmas 'This is our harvest time, make it pay.'

The consumer mechanics revolution began with inventions

like Latham Sholes' typewriter of 1868 which allowed Mark Twain to lay claim to the first novel printed from a typescript with *The Adventures of Tom Sawyer*. By 1871, 700,000 sewing machines a year were being made in America, most of them by Isaac Singer. Alexander Graham Bell obtained his telephone patent in 1876 and three years later the *New York Herald* newspaper gave front page treatment to Thomas Edison's first incandescent electric lightbulb. George Eastman put his Kodak camera on the market in 1888, ushering in the era of family snapshots and tangible memories. In the 1890s Edison laid the foundations for Hollywood and a vast entertainment industry with the invention of a practical movie camera, 35mm celluloid film and a phonograph. Emile Berliner and Eldridge Johnson perfected the Gramophone and shellac records to go with it by 1897.[106]

Small arms were crucial to the development of consumer durables because Eli Whitney developed the Uniformity System for manufacturing muskets (on a 1798 contract to John Adams' Government), whereby each piece was made 'independently and in large quantities' and was interchangeable. By 1809, after a decade of trial and error, 'Whitney had finally proved and successfully applied the basic idea of American mass production, without which the American Standard of Living would have been inconceivable.'[107]

Military contracts have long been recognized as an important driving force behind both employment and technological change. Just as muskets were the first fruits of true mass production, and thereby the forebears of our consumer society, so was perfection of the motor car engine a product of the armaments machine. Techniques which had been developed to standardize the boring of gun barrels were used in the production of engine cylinders.

Apart from the arms trade, the oldest of the modern consumer industries was the making of clocks and watches. Clock manufacture, which began as a craft tradition, soon moved towards mass production as demand grew. Europe was producing about 400,000 watches a year by the late eighteenth century, with half of these being made in Britain. A century later, output had more than quadrupled, with the Swiss controlling 75 percent of production. By 1914, the Swiss were making approximately 17 million watches every year.[108]

The rapid expansion of manufacturing industries during the nineteenth century was accompanied by an explosion of occupational disease. The demand for consumer goods was on the rise, and in their haste to turn a quick profit from selling china, cutlery, linoleum, Christmas decorations, mirrors, textiles and a host of other commodities, factory owners ignored the health impacts of the industrial pollution their employees were suffering daily inside the workshops.

Potter's rot, brassfounder's ague, soot wart and flock fever were all common. Cutlery grinders and scissormakers suffered terrible breathing problems caused by inhaling metal filings; flax workers got asthma from cotton dust and pottery workers constantly inhaled tiny particles of clay, flint and felspar. Hat and mirrormakers went mad from mercury poisoning, white phosphorous rotted matchmakers' jaws, confectionery workers and wallpaper tinters were prone to arsenic-induced skin diseases, and printers and glaziers were stricken with plumbism.[109] Millions of workers in Europe suffered debilitating industrial diseases caused by the pollution of their working environments with dusts and toxins during the industrial revolution.

From Bike to Buick

Promoted today by environmentalists as an alternative to driving a car, the bicycle was actually the forerunner of the car. Peugeot, Opel, Humber and Rover, for instance, all manufactured bicycles before they turned to automobiles. A combination of aggressive marketing and fashion turned the bicycle into a bestseller. At the peak of the pre-car British bicycle market, people were buying a quarter as many cycles as were sold in 1990. By the outbreak of the First World War, there were probably more than nine million bicycle owners in Europe and the United States. In the words of Goodman and Honeyman 'the bicycle had become a pre-eminent mass-produced consumer durable'.[110]

It was from this thriving industry, with its innovations in

machine tool design, that automobile production grew. By 1913, Henry Ford was producing nearly 200,000 Model Ts a year. A couple of generations later, an advertising campaign by the environmental group Greenpeace turned Ford's own phrase 'any colour so long as it's black' against his company by referring to the colour that vehicle exhaust fumes turn human lungs.

The automobile is a schizophrenic machine. It has been a bringer of freedom and of pollution; helping to define both individuality and equality, and has become emblematic of society's aspirations and of its failures. It is the ultimate interface between consumption and pollution.

In the beginning the car promoted the separation between the rich and the rest or, as the philosopher Andre Gorz put it, 'until the turn of the century, the elite did not travel at a different speed from the people. The motorcar was going to change all that. For the first time class differences were to be extended to speed and to the means of transportation.'[111] This was truer of Europe than it was of America. Daniel Boorstin, describing the westward migration of the early nineteenth century observed, 'It was most important to arrive there first, or, if not first, then as soon as possible. It was haste that decisively shaped the technology of American travel.'[112] Haste may have shaped it, but mass production and marketing 'know-how' enabled every American to aspire to owning a car. By 1921 ten million of them did, and ownership tripled over the next decade.[113]

Innovation after innovation allowed the spread of the car in the United States. The development of the assembly line in Ford's Detroit works was a first. So too, according to Witold Rybczynski, was the fact that Ford embraced the five day working week in 1926, amidst a storm of criticism from other industrialists. According to Rybczynski, Ford's 'rationale was that an increase in leisure time would support an increase in consumer spending, not least on automobiles and automobile travel'.[114]

The high pressure marketing and consumer credit booms of the 1920s and 30s also played their parts, as did less well publicized initiatives. In the 1920s, for example, the introduction by General Motors (GM) of the annual model change primed the car market to exploit the consumer's desire to own a different, better, more modern car than his neighbour. And then in 1936, along with Standard Oil and the Firestone Tyre Company, GM started National City Lines, a company whose purpose was to buy up and close down rail and tramways that might compete with the car as means of transport.

Gorz was on target when he said 'Unlike all previous owners of a means of locomotion, the motorist's relationship to his or her vehicle was to be that of user and consumer – and not owner and master. This vehicle...would oblige the 'owner' to consume and use a host of commercial services and industrial products that could only be provided by some third party.'[115] In less than 50 years the car had formed the bridge from the last stages of western industrialization to the consumer boom of the second half of our century.

In 1909, in the spirit of his age, Henry Ford said 'I'm not creating a social problem at all. I'm going to democratize the automobile. When I'm through everybody will be able to afford one, and about everyone will have one.'[116] On the first point Ford could not have been more wrong and on the second hardly more accurate for, at least in the USA, there are two cars on the road for every three living Americans.

Consumption and Population

The flip-side of the consumption coin is population. If everyone in the world were to own a car, there would be 5.5 billion of them on our roads, resulting in global gridlock and unsupportable air pollution. As it is there only about 500 million cars, but the number is growing at an alarming rate. Ask a car company representative what is the limit to the number of cars in the world and you will be answered with silence. Today's commerce cannot think in terms of limited markets, only the new untapped markets of Asia, Africa and the former Soviet Block. What is true for cars is also true for cigarettes, TVs, Coca Cola and every other icon of the consumer age. Our consumption patterns, the lifestyles of the rich and shortsighted, cannot be cloned around the world without finally overwhelming the planet's ability to repair environmental damage.

Carrying capacity has been defined as 'the maximum impact

that the planet or any particular ecosystem can sustain'.[117] Not enough is known about the limits to carrying capacity, but it is not infinite, and it is clear that the more people there are on the earth, and the more they consume, then the greater will be their impact on the environment. Ralph Waldo Emerson realized that man alone among species would threaten to overwhelm the carrying capacity of the earth. In 1868 he told a Boston audience 'that a pot of earth may remain a hundred years the same; but put in a seed, and all is changed – not the seed only, but every atom of earth. Now, put a man into the world, and see how soon that great pot will be changed.'[118]

Emerson's pot of earth is changing, and the world is already full of humans. From the birth of Christ it took more than 1600 years for world population to double to around 600 million. Since then, with exponential growth, the population has doubled more than three times. We have reached 5.5 billion people and this is expected to double once again within a century. Three babies are born every second, 250,000 each day, roughly 95 million every year. About 90 percent of the coming growth will occur in the developing nations.[119]

As a result of the consumer revolution described above, the 21 percent of the population that lives in the industrialized world currently consumes roughly 80 percent of global resources and is responsible for the vast majority of waste and pollution. Even now it would not be possible for the earth to survive global consumption patterns as profligate as those current in Europe or North America. The link between population and consumption is not a simple equation though. Not every individual has the opportunity to consume the same, and not every form of consumption is comparable. Energy used to heat a house or cook a meal in Africa cannot fairly be compared with energy used to overheat a house or to drive a lawnmower in America.

Growth in numbers is not the only demographic pressure affecting the environment. The distribution of that growth, increased urbanization, and some forms of migration are all also of great importance.[120] In the past these factors have all played their part in environmental and cultural destruction. Pollution itself played a part in limiting the growth of previous human populations and civilizations, through the spread of disease in particular. Famine, climate, geography and war have also been responsible for fluctuations in the past. Fernand Braudel has suggested that France was already overpopulated in terms of its resource base by the year 1600, and that this resulted in massive migration to Spain.[121]

The Spanish themselves sent only a small number of people to the Americas, but these people were responsible for the deaths of millions of Indians from hundreds of tribes. The same happened to native peoples across the globe as the steadily growing European population sought out new lands and new resources to feed their land hunger and new lifestyles. Australian aborigines, native Americans, Caribs, Aztecs, Maoris, Hottentots; all fell victim to the 'civilizing' sweep of colonial powers. These population shifts and reductions changed the face of the earth, as did the enforced slavery of millions of Africans who were transported to fuel the agricultural engines of America and the West Indies.

Population movement continues to stress the environment and local cultures today. Landless peasants slash and burn into the jungles of the Amazon, refugees from war, famine and oppression stretch the resources of a dozen African countries, and hotels catering to millions of foreign tourists spew sewage into the Mediterranean, Caribbean and South China Seas.

Population growth itself can be a threat even to the best managed and protected ecosystems. Just one example can be given by Lake Nakuru in Kenya. For over 30 years a focus of successful international conservation efforts in Kenya and famous as a home to millions of flamingos, Nakuru is under severe threat once again. Traditional conservation methods and the creation of a well-managed reserve (with a solar powered electric fence to protect rhinos from poaching) are no match for the problems caused by population growth.

The lake upon which flamingos are so numerous that, from a distance, they appear as a pink slick around the shore, receives all the sewage from the rapidly growing town nearby and masses of run-off of soil and chemicals from intensifying agriculture in the surrounding landscape. The population in the surrounding area is rising fast at over 6 percent a year.[122] Growing human needs require that 25 million gallons of water are extracted daily from rivers and boreholes that would otherwise feed the lake. In return, 15 million gallons of sewage, agricultural run-off and polluted waste water reach the lake every day. Without action to improve sewage treatment facilities and adopt agricultural strategies in the short term, as well as stabilizing population in the longer term, no amount of fences can save Nakuru from the people problem.

Another example of how the twin threats of population pressure and over-consumption can negatively affect wildlife may be illustrated by the plight of the beautiful monarch butterfly. This extraordinary insect migrates south from its summer breeding grounds in Canada and the USA to spend the winter in huge colonies in the mountain Oyamel pine forest of Mexico. Known to locals for generations, but not discovered by scientists until the mid-1970s, these wintering sites present one of the most spectacular sights in nature. Huge numbers of butter-flies – sometimes 10 million to a hectare – cluster like drifts of orange snow on every available pine branch. A constant fluttering sound fills the forest and occasionally the weight of the wintering insects causes a branch to snap and crash to the ground in an explosion of bright wings.

Despite the fact that the Mexican government has designated most of the monarch wintering sites as protected areas, the relentless pressure of poverty and population growth in the surrounding areas is eating away at the forest. In the absence of alternatives, local people depend on wood from the forest for fuel and income. According to scientist Lincoln Brower, who has been studying the ecology of the monarchs in Mexico for more than 15 years, at least one of the best wintering sites has already been virtually destroyed by timber cutting. Couples may have had 12–14 children in some of the villages close to the handful of butterfly reserves that remain. Not even protected areas status can prevent the slow decline of the forests while wood provides the subsistence needs for its fast-growning population.

Unfortunately for the monarch, it is no longer safe even when it moves north across the Mexican border in the spring. Brower believes that the intensification of agriculture in the consumer culture of North America will help to speed the demise of this fragile insect. Increasing use of herbicides may be harming milkweed, the monarch's only food plant, and the build up of pesticides in the insects' bodies is likely to inhibit breeding success. Clearly, the monarch butterfly is caught between the fast-closing pincers of population and consumption at both ends of its migratory range. It has little hope of survival unless these pressures can be relieved.

With modern technologies, increased populations and ever more scarce supplies of natural resources, it is sobering to think how small the world has become in recent years. Where Columbus took four months to cross the Atlantic, we take just a few hours. Geographic scale changed no less than consumption patterns during the last two centuries. For example, less than a day's drive from Lake Nakuru, tourists from all over the world now flock daily to the game parks at the foot of Kilimanjaro. Yet in the recent past this was a mountain so remote, and thought so strange (due to the accounts of its snow-capped peak) when first discovered by Europeans in 1848, that it took thirteen years to even prove its existence.[123] Until the beginning of the nineteenth century, most of Africa was unknown to outsiders. The era of European exploration and colonial expansion in Africa that made the names of men such as Mungo Park, Livingstone, Burton and Stanley famous, coincided not only with great scientific advances and a revolution in consumption patterns but also with rapidly expanding populations in Europe.

Population pressure became a matter for debate with the publication in 1798 of Thomas Malthus' *Essay on the Principle of Population*. Malthus presaged today's concept of 'carrying capacity' when he argued that there were natural laws that would set limits to the growth of human population. Then came Charles Lyell with his theories of geological time measured in millions of years, and Charles Darwin with the *Origin of Species*. As the scientists and philosophers sought to explain man's relationship to nature, English commentators began to link population growth with social advancement. An editorial in the *Illustrated London News* in 1859 exclaimed that 'At no period did invention make more rapid strides, were the arts more improved, was science more swiftly extended, or humanity more cultivated, than within the last few years, while population has been rapidly increasing and been unprecedentedly great.'[124]

With hindsight, we can see that although progress can be associated with population growth, it may be massively outweighed by multiplied pollution and environmental degradation. The first consumer revolution provided the means for a greater proportion of humanity to live better than it had before, but it also opened the door to over-exploitation and waste of resources. Increased consumption combined with population growth can be a potent mix.

Chapter Five

WATER POLLUTION AND CHEMICAL CONTAMINATION

*Verily, a muddy stream is man. One must be a sea
to be able to receive a muddy stream without
becoming unclean.*

Friedrich Nietzsche

Not so Pure

Water is the life-giver and the hydrological cycle is central to human existence. As surely as rain falls on the land, so too must much of it drain into the oceans. People have understood the importance of water for millenia, and yet it is one of the most abused and exploited substances on our planet. Adam Smith noted 'Nothing is more useful than water; but it will purchase scarce anything; scarce anything can be had in exchange for it.'[125]

In 1977, the UN Water Conference in La Plata declared that 'all peoples have the right to have access to drinking water in quantities and of a quality equal to their basic need'.[126] Despite the fact that there is more than enough water on the planet to meet this objective, water pollution and scarcity have been throughout history, and remain today, one of humankind's most intractable problems.

The politics of clean water really took off in 1854, after John Snow demonstrated a clear link between sewage leaking into London's Broad Street well and cholera outbreaks in the city.

52

Eventually, better management of municipal sewage, sand filtration techniques and water chlorination followed this discovery, making drinking water supplies safe in most developed countries by the early twentieth century. In these countries the new freshwater pollution problems of oxygen depletion, eutrophication, heavy metal contamination, acidification, toxic chemicals and nitrates began to be manifested from the 1950s onwards.[127]

While these problems are all, to one degree or other, present also in developing countries, it is human sewage which remains the most important pollutant of their freshwaters and coastal zones. A World Health Organization report published in 1992 found waterborne infections (such as cholera, diarrhoeal diseases, typhoid and poliomyelitis) to be the 'largest single category of communicable diseases contributing to infant mortality in developing countries.'[128] These diseases kill four million children and a million adults every year. Schistosomiasis or bilharzia, a parasitic disease of the bladder and intestines spread by pollution of watercourses by human excreta, affects a further 200 million people in the tropics.[129] Worldwide, nearly a billion people lack access to safe and adequate water supplies, and at least 1.7 billion lack adequate sanitation.

The situation has not markedly improved during the last half century. For example, a 1928 report in the scientific journal *Nature*[130] described the situation in the Indian province of Madras:

> *No system of drainage is in practice, with the result that pools form in every depression during the rainy season and stagnate in the hot weather. Of sanitary arrangements there are almost none so that the soil in and around the village becomes polluted and all waterways are a positive danger.... Little wonder, then, that the deaths from preventable diseases reach appalling figures.*

Even today, less than 10 percent of India's towns and cities have even partial sewerage and effluent treatment. And the WHO said of Madras 64 years after the *Nature* report, that less than a third of the city is served by a sewerage system and 'raw sewage flows freely into the metropolitan area's natural watercourses at many points'.[131]

Nevertheless, there are some success stories. In 1986 Rajiv Ghandi launched the $140 million Ganges Action Plan aimed at cleaning the full 2500km length of the sacred river. Apart from the torrent of agrochemicals, industrial waste and sewage that daily flows into it, the Ganges suffers from the unsightly menace of up to 30,000 partly cremated Hindu corpses each year, sent on their journey to final salvation. In an inspired piece of natural waste management, more than 24,000 carnivorous snapping turtles have been released into the holy waters to scavenge on floating body parts.[132]

The pattern of uncontrolled sewage pollution is repeated throughout the industrializing world, and is compounded by the impacts of agricultural intensification (causing high levels of pesticide and fertilizer run-off to reach surface waters) and from industries such as tanneries, mines and petrochemical works whose effluents are often pumped directly into rivers or coastal waters without restriction or rebuff.

Just one of many industries causing water pollution, the pulp and paper sector has become one of the most controversial. Growing consumer demand for paper products that are recycled, use less energy in manufacture and don't cause pollution have forced the industry onto the defensive. Approximately 240 million tons of new paper are produced every year, requiring the felling of up to four billion trees, often from old-growth forests in countries such as the USA, Russia and Indonesia. By far the worst by-product is the one billion tons of organochlorine effluent that flows into the world's rivers from pulp mills every year.[133] More than a tenth of global chlorine output is used in the paper industry, mainly in a bleaching process which has been linked to pollution of rivers by dioxins. These chemicals are some of the most toxic compounds known to man, and their threat to human health and wildlife have led environmentalists to call for a ban on paper whitened by chlorine bleaching.

From the River to the Sea

The great deep, or 'Mother sea' as Rachel Carson called it, from which 'the first pioneer of land life crept out on the shore' some 350 million years ago, is in deep trouble.[134] We've come a long way since Silurian times, most notably since the advent of *Homo sapiens* on the planet. The great primeval ocean was the cradle of life on earth, but more than two billion years after scientists believe that life began, people can no more control the marine environment than when they first stood at its edge in the prehistoric past.

We cannot control the sea, but we can abuse it. Joseph Conrad mourned the end of the age of sail when he observed 'The machinery, the steel, the fire, the steam have stepped in between man and the sea. A modern fleet of ships does not so much make use of the sea as exploit a highway.'[135] Since then, we have not just used the sea as a means of transport, but have regarded it as the ideal dumping place for our detritus and waste. And today's fleet on Conrad's highway has among it ships that dump sewage sludge, incinerate toxic waste, and transport chemicals, oil and plutonium.

The sea has always been an attractive option for the disposal of waste. Long the dumping ground of choice, it has received our liquids and our solids; thrown overboard from ships, shovelled down cliffs, piped into rivers, canals and ditches and even blown from our chimneys by the four winds. Even the landlocked nations share the blame. Effluent from the Swiss vineyards flows down the Rhone to the Mediterranean, and from Hungarian chemical works down the Danube to the Black Sea.

It is virtually impossible to stand on the deck of a ship anywhere in the world's oceans and fail to spot, after a short while, some flotsam or jetsam. A small oil slick from a tanker's bilges, black plastic garbage bags flung from a luxury cruiser, a few empty oil drums or chemical containers washed overboard in a storm, tropical logs from a cargo boat supplying the European DIY market, or the inevitable floating armchair or settee originating Heaven only knows where. To test the all-pervading diffusive power of modern pollution, walk a kilometre of shoreline anywhere outside Antarctica and you will find some combination of foreign plastic bottles, globs of sticky tar, tangled nylon fishing line, recently poisoned seabirds, ulcerated fish, broken glass, rusty Coke cans and soggy cigarette filters. These testimonials to our power to kill and consume are merely the most visible signs of the deterioration of the seas.

Much early sewage treatment relied on the ability of the oceans to dilute discharges and subsequently render them harmless. This worked for New York, for instance, until the 1930s, when the huge quantities of sewage being dumped at the mouth of the Hudson river overwhelmed the Atlantic's cleansing capacity, and a lifeless area of foul, black sludge spread over the sea bed. On the other side of the 'big pond', Britain alone among the North Sea states has refused to stop dumping sewage sludge at sea.

One unusual example of marine pollution dilution was demonstrated by the Chinese imperial commissioner Lin Zexu who had been sent to Canton to wipe out the opium trade. In 1839, 20,000 chests of opium confiscated from British traders were adulterated with salt and lime and then dumped into an ebbing tidal creek. Lin consigned the narcotic cargo to the spirit of the Southern Sea with a prayer: 'you who wash away all stains and cleanse all impurities...poison has been allowed to creep in unchecked till at last barbarian smoke fills the market...tell the creatures of the water to move away for some time to avoid being contaminated.'[136] This episode hints at a solid Chinese understanding both of the concept of assimilative capacity of the sea, and of the possible impacts of pollution on marine life.

Once in the sea, pollutants cannot easily be controlled or removed. In many cases they cannot even be traced or tracked. Fifty percent of the world's population lives within 60 kilometres of the coast,[137] tens of millions of people depend on marine life for the bulk of their protein intake, and yet most people look no further than the surface of the sea, the parameters of their daily lives being set by their landward surroundings. Even marine biologists and oceanographers still know little about the sea and less about the mechanisms or impacts of its contamination.

But advances in fish biology may eventually help biologists to gauge the threat from pollution. Cancers in fish are becoming epidemic in highly polluted waters such as Europe's North Sea and Ohio's Black River. Scientist John Harshbarger of the Smithsonian Institute in Washington DC has been recording tumours in fish from all over the USA for more than 20 years. He has suggested that the famous naturalist Henry David

Thoreau was the first to find cancers in fish when he described black felt-like growths on pout caught near Walden Pond. Harshbarger found the same growths on fish there more than a century later, and identified them as melanomas. He believes that certain types of cancers, particularly liver cancers, may provide some of the most potent and reliable indicators of water pollution threats to human health. The canary in the coal mine may become the goldfish in the washbasin.

Pollutants in the sea are chemically transformed, remain in the water or are bonded with sediment; they can accumulate in the food chain and are transported around the globe on the currents. These poisons which we have long since said goodbye to as they disappeared up a flue stack or down an effluent chute come back to us in our dishes of gourmet seafood, in refreshing rain showers and even in sea spray as our children play on the beach.

The pressures on the sea are growing. The devastation of world fishery stocks, along with the mismanagement and development of most of the planet's coastal zones combine with pollution to present a fearful assault on the health of the oceans. One of the fastest growing threats is tourism. Approximately 100 million tourists visit the Caribbean basin every year[138] and more than 110 million, the Mediterranean coast.[139] These transient visitors stretch the waste management and sewage disposal capabilities of thousands of small coastal towns and villages beyond their maximum capacity. Mauritius, for example, famous for the extinction of the dodo in the 1670s, already had laws against pollution by effluent from indigo factories and sugar mills in 1791.[140] But in 1991 the tiny island hosted 300,000 visitors, mainly from South Africa and Europe. Their untreated sewage is being dumped in the lagoons, and the marine life they come to see is suffering from their presence.[141]

The phenomenal growth of tourism in some developing countries is likely to put a severe strain on coastal ecosystems. Marine turtle nesting sites in Malaysia, for instance, could become the victims of the recent success of that country in attracting foreign visitors. Malaysian tourism grew by a staggering 54 percent in a single year from 1989. The 7.5 million visitors in 1990 helped tourism leapfrog logging, rubber and palm oil to become the third largest foreign exchange earner after manufacturing and petroleum.[142]

Coral reefs too are impacted by tourists. Sewage from tourism developments and from fertilizer run-off can cause corals to be smothered by abnormal growths of seaweeds and oxygen to be depleted by phytoplankton blooms. Impacts can already be seen on Florida's declining reefs. Physical damage caused by tourists trampling on corals, diving and dragging their boat anchors through reefs is also increasingly common in places like Israel, Kenya and the Maldives.[143]

Tourist pollution is just one source of so-called 'land-based' pollution. As a whole, this sector accounts for 77 percent of the total marine contaminant load.[144] Of this, a little over half enters the sea from the atmosphere, and for some pollutants, such as lead (which comes primarily from vehicle exhausts), zinc and cadmium, the atmosphere is the main pathway to the sea. Approximately 80–99 percent of PCBs, DDT and HCB (hexachlorobenzene) in the open oceans come from the air, often having travelled thousands of kilometres.

Another crucial source of pollution is the agricultural sector. Its impacts are as heavy on freshwater systems as they are in the sea, the main threats being run-off and leaching of pesticides, nitrates and phosphates. The World Bank estimates that 550 tons of pesticide reach the Mediterranean sea alone every year.[145] Nutrients arriving in the Baltic from the air and via rivers are contributing to serious eutrophication in this semi-enclosed sea, and the North Sea is plagued with an annual input of 1.5 million tonnes of nitrogen (approximately 40 percent of this from the air due to vehicle exhausts and power station flue gases) and 100,000 tonnes of phosphorus.[146] The sources of the phosphate are agricultural fertilizer run-off and domestic waste water containing detergents.

This massive input of chemical nutrients stimulates the growth of marine algae, just as it does agricultural productivity. In water, however, the over-nutrition leads to blooms of algae and the consequent death of local aquatic life as decomposition of the short-lived plant matter soaks up most of the available oxygen. Some of these blooms can also have a directly toxic effect, as with the 'red tides' caused by phytoplankton, and first recorded off the coast of Florida in 1916. Now an annual occurence in the Gulf of Mexico and common off the Japanese and Italian coasts, as well as many other seas polluted by sewage and agricultural nutrients,[147] red tides can cause serious skin irritation in swimmers, or death through consumption of infected shellfish.

Oil and Plastics

The pollutant most people first associate with the sea is oil. Although widespread and common, it is predominantly a problem for seabirds and amenity beaches. Accidental oil spills have, however, elevated oil to mythical status as a pollutant. The *Torrey Canyon* disaster in 1967, when 119,000 tonnes of crude oil polluted miles of the French and British coasts was one of the first 'celebrity' oil spills. It was followed by the *Amoco Cadiz* (220,000 tonnes off the British coast in 1978) and, most recently, by the *Exxon Valdez* which grounded in Prince William Sound and spread crude oil over 2600 square kilometres of Alaskan coast and waters. As well as tens of thousands of seabirds, more than 1000 rare sea otters died in the oily wake of the *Exxon Valdez* catastrophe.[148]

Public concern over oil spills seems to be primarily driven by the newsworthiness of the events and associated with dramatic TV footage of shiny, slimy, oiled birds and wildlife dying a painful death. This is compounded by the apparent indifference of the big multinational companies whose produce is vividly shown by the media to be blackly blanketing the coastline. The public's view of the oil companies is summed up by a Sierra Club T-shirt slogan seen in the aftermath of the *Exxon Valdez* disaster – 'We don't care, we don't have to, we're Exxon'.

On top of the tanker accidents come the other oil disasters. The blowout of the Mexican exploration well, Ixtoc 1, in the Gulf of Campeche in 1979, for many years held the distinction of causing the largest spill. The well pumped out 20,000 to 40,000 barrels a day until it was plugged, and devastated crab and mollusc fisheries over hundreds of square kilometres. Then came the Gulf War whose burning oil fires turned day into night and which spawned apocalyptic visions of global cooling because of the smoke. The global cooling never happened, and it is sobering to note that the total emissions from the months of burning wells contributed a mere 1 percent of total fossil fuel emissions for the globe in 1991.[149] Between us, with our industry, our cars and our electrical appliances we multiply the pollution impacts of the Gulf conflagration a hundredfold.

But the impacts of the Gulf War are not to be played down. The release of nearly a million cubic metres of oil (roughly twice the amount of the Ixtoc 1 blowout and 20 times the *Exxon Valdez*) by Iraqi soldiers, into the Gulf from the Sea Island terminal caused havoc to the Saudi coastline. Surveys of the local fisheries have shown that post-war shrimp populations in the north western Gulf may have collapsed to just 10 percent of 1989 levels, threatening severe social and economic hardship for the coastal communities that depend on these stocks.[150]

Despite these high profile oil events and the publicity they receive, only about 12 percent of marine oil pollution results from tanker accidents and the number of major oil spills world-wide dropped by almost 75 percent between 1974 and 1986. Transportation is a major source, however, and the biggest category of oil release comprises deliberate and operational discharges of crude, bilge and fuel oils.[151] More than a third of marine oil pollution originates on the land, from industrial and municipal waste, and much of this is more toxic than the crude oil released in tanker spills. Lubricating oils contaminated with poisonous non-degradable chemicals such as PCBs are commonly released into the coastal zone.[152] American car-owners alone pour ten times more motor oil down their drains and sewers every year than the amount of crude oil that was spilled by the *Exxon Valdez*.

Plastics too are of major concern for the marine environment. Invented in the 1860s and not widely available until the early decades of this century, plastics of one sort or another now pervade our everyday life more thoroughly than almost any other material. Their energy and resource cost is high too: in the Netherlands, for instance, plastics account for 2 percent of total energy use and 4 percent of oil consumption.[153]

The dumped and accidentally spilled pellets that are the raw material of the plastic manufacturing industry are a deathly lure to seabirds which ingest them, and the plastic rings from six-packs of beer lodge around the necks of countless gulls. Floating plastic bags are a significant cause of mortality amongst sea turtles which mistake them for jellyfish and eat them. Discarded and lost plastic fishing tackle and nets account for much of the annual kill of marine mammals. North Pacific Dall's porpoises, fur seals in the Aleutian Islands and Hawaiian monk seals are all reported to be threatened by entanglement in nets.[154] The curse of drift nets – hanging nets that can be 50 km long, left to float submerged in mid-ocean – account for the deaths of millions of non-target fish, including sharks, along with whales, dolphins and seabirds every year. The UN reported that in the 1988–89 season alone, between 300,000 and 1 million cetaceans, mainly dolphins, were killed in these submerged death nets.[155]

Toxic Chemicals

The World Health Organization recently confirmed that although most information on chemical hazards has been gathered in the North, most of the exposure and impacts are in the South.[156] The industrialized nations dominate the chemicals trade. The EC is the world's largest producer: over 100,000 chemical substances – 20,000 of them classified as dangerous – are available for sale in more than 1 million products.[157] The toxic residues from the manufacture and use of these products are increasingly finding their way into rivers, lakes and ground-water and also coastal waters.

Most notorious among the early tragedies caused by chemical pollution was the painful death and disability that afflicted thousands of people living around Japan's Minamata Bay in the 1950s and 1960s. The victims ate shellfish and fish contaminated by methyl mercury. And the Chisso company, which had been discharging mercury wastes from the manufacture of acetaldehyde, is still paying compensation to the victims.[158]

The massive Great Lakes Basin of North America was another early chemical pollution victim. Described as 'sweet-water seas' by the early explorers, pollution had become a problem for these waters at the beginning of the twentieth century. By the 1960s the Cuyahoga River in Cleveland, Ohio, which flowed into Lake Erie 'ran a chocolate brown or rust colour and was choked with debris, oils, scum and floating organic sludges'. The situation was so extreme in 1969 that the river caught fire and burned bridges crossing it in an inferno of toxic flame and fumes.[159] Although the acute pollution problems of the Great Lakes are receding, the danger now lies with chronic contamination by more than 350 toxic substances. Efforts at clean up are concentrating mainly on the 11 worst ones. This critical list includes PCBs from the electrical industry, dieldrin and toxaphene from agriculture and mercury from the metallurgical, chloralkali and paint industries.

New scientific research is fuelling fears that chemicals such as these in the environment of the Great Lakes and elsewhere may pose a much greater health risk than previously thought. Several studies now suggest that dozens of common chemical contaminants may be wreaking havoc with the reproductive systems of wildlife and humans. These chemicals appear to have characteristics that resemble the human sex hormone oestrogen, and their presences may trick the body into altering its biochemical responses. Scientists believe that the results of this chemical confusion include lowered sperm counts, undersized penises and testicular cancer in men, and breast cancer and infertililty in women.[160]

New evidence for the links between chemical pollution and reproductive defects is emerging from human studies in places as diverse as Michigan, New York's Long Island, Germany and Taiwan. Supporting evidence also comes from a string of wildlife impact assessments. Alligators, turtles, birds, shellfish, mammals and fish with high levels of endocrine-disrupting chemicals that are associated with these problems have all demonstrated reproductive disorders. The list of common chemicals that are associated with these problems now stretches to more than 42, including pesticides such as Atrazine, Zineb, DDT, Lindane, Mirex and Aldicarb, and others such as lead, PCBs and styrenes.[161] Some of these chemicals are no longer in common use, but will continue to haunt us for generations while they remain in soils and sediments and slowly escape into the food chain.

Banned chemicals such as the fire-retardent Mirex still pollute the lakes' sediments and fish because of past discharges. Mirex from old spills near Niagara and the Oswego River travels nearly 1000 kilometres in contaminated eels which migrate from Lake Ontario to the St Lawrence estuary. In these coastal waters, the 'most polluted mammals on earth', the local beluga whales, eat an average of 88 kilos of eels a year. Mirex and other toxic contaminants are devastating the whale populations. Dead whales that end up on the shore often have to be treated as toxic waste and disposed of accordingly. The pesticide DDT, highly restricted for use in the US, is still manufactured there and sold for mosquito control and fruit growing in the tropics. Residues from DDT spraying in Mexico are being returned by the winds as atmospheric pollution to the Great Lakes, where America's national symbol, the bald eagle, ingests them in fish and consequently has trouble breeding successfully.[162]

The issue of the sale of chemicals banned in the country of manufacture for use in another, usually developing, nation is a highly sensitive one for the chemical industry. And there is also the associated question of the transfer of manufacturing to developed countries where environmental standards are less strict. These are the 'double standards' issues, and the multinational corporations have been most adept at exploiting the differences in national legislation. There have been numerous documented cases of Northern multinationals' subsidiaries operating double standards leading to pollution in the South. Examples include Mitsubishi and the dumping of radioactive thorium waste in Malaysia; pollution from a Bayer chromium plant near Mexico City; water pollution from the manufacture of dyes in Bombay from a factory partially owned by the Italian Montedison group; and mercury pollution in Nicaragua's Lake Managua from the US Pennwalt Corporation's chlorine plant.[163]

It was the explosion at Union Carbide's pesticides factory in Bhopal in 1984 that blew the double standards debate onto the front pages of newspapers around the world. The final death toll will probably never be known, but somewhere between 2000 and 5000 people died and nearly a quarter of a million suffered, chiefly from damage to lungs and eyes. The cloud of vaporized methyl isocyanate that hung over Bhopal on that windless

63

December night, bringing tortured pain to tens of thousands of slum dwellers, had formed after 40 tons of the chemical accidentally heated up in an improperly maintained storage tank. Union Carbide at first denied, and then later admitted, that its subsidiary's plant in Bhopal was not complying with the same safety standards as its sister factory in Institute, West Virginia.[164] The horrific events at Bhopal may have changed attitudes to multinationals operating in developing countries, but it hasn't changed the fact that producing chemicals is a risky business wherever it's done. Leaks and explosions in the US petrochemicals industry killed 159 people and injured at least 2200 more in a six year period ending in 1992.[165]

Since 1901, when the Spindletop oil well was opened in Beaumont, Texas, that part of the USA has been home to one of the greatest concentrations of oil refineries and chemical plants in the world. The Galveston Bay port of Texas City is home to seven major petrochemical facilities, employing roughly one quarter of its citizens and providing 75 percent of its tax revenue. It was here that America's worst ever industrial disaster occurred in 1947 when two freight trains loaded with explosive ammonium nitrate fertilizer crashed and caught fire, killing 576 people.

Four decades later, accidents will still happen. Although no one was killed, one of the latest Texas City disasters happened when an error by a crane driver at the Marathon Oil refinery caused thousands of gallons of acid vapour to shoot into the air. This was the largest ever release of hydrofluoric acid, a catalyst for oil refining. The resulting 'choking chemical cloud' caused 3000 people to be evacuated and 1000 people to submit for medical attention.[166]

The fear of chemical accidents is becoming ingrained in urban humanity. Few people in the industrialized world have not witnessed at first hand or on TV, the devastating consequences of some kind of chemical explosion, in war or in peacetime. The threat of immediate death is now moderated by the greater fear of long term, unknowable effects such as cancer. Don DeLillo, one of the great writers on the neuroses of the consumer society, constructed the whole of his 1986 novel *White Noise* around a chemical accident.

After seeing the 'black billowing cloud, the airborne toxic event' which 'moved like some death ship in a Norse legend', DeLillo's character Jack Gladney muses:[167]

This was death made in the laboratory, defined and measurable, but we thought of it at the time in a simple and primitive way, as some seasonal perversity of the earth like a flood or tornado, something not subject to our control. Our helplessness did not seem compatible with the idea of a man-made event.

Chapter Six

THE POISONED ATMOSPHERE

What will my son breathe?
Mashed shit.
Carbonic gas.
Metallic dust.
And all of it at an altitude of about one
and a half miles, crushed under a
layer of frozen air, and surrounded
by a jail of circular mountains:
garbage imprisoned.

Carlos Fuentes

Acid Rain and Air Pollution

'Acid rain falls on the just and the unjust and also and equally on the rich and the poor' said John Kenneth Galbraith.[168] His point was that in the late twentieth century, the emissions which cause acid rain are transported far from their source, and the entire population of the polluted region is exposed to the consequences.

It has not always been thus. Although air pollution had certainly had its impact on the richer classes during the industrialization process of the late eighteenth century, and throughout the nineteenth, it was primarily the poor who were affected. As Engels observed of Manchester in 1845 'These east and northeast sides...are the only ones on which the bourgeoisie has not built, because ten or eleven months of the year the west and south-west wind drives the smoke of all the factories hither, and that, the working-people alone may breath.'[169] A combination of smoke, toxic fumes, unsanitary and cramped living conditions conspired to radically depress life expectancy in many urban slum areas in Britain. An 1844 Royal Commission found that in some parts of Nottingham, the mean age of death was as low as 18 or 19.[170]

The main sources of air pollution until the middle of the nineteenth century were the domestic burning of fuel, and the production of iron, steel and chemicals. It was this latter industry that first succumbed to air pollution legislation. Reports of chemical devastation of the countryside and searing of crops led the gentry to fear for the value of their land, and the House of Lords Select Committee on Injury from Noxious Vapours produced a report that led to the Alkali Acts of 1863.[171] These pioneering laws set out to cut emissions of hydrochloric acid from the manufacture of the main chemical products of the day, which were salt-cake (sodium sulphate) and soda-ash (sodium carbonate), by 95 percent.[172]

The man entrusted with the task of enforcing the legislation was Britain's Alkali Inspector, Angus Smith, who in 1872 was to publish his book *Air and Rain*. This seminal work clearly showed that levels of sulphate were an order of magnitude higher in industrial towns and cities than they were in the countryside. Smith coined the phrase 'acid rain', and air pollution became a public issue with many articles in the press. One such piece in The *Illustrated London News* of March 1872 blamed acid pollution for the death of vegetation around Newcastle and Liverpool and lamented that 'The erection of very high chimney shafts were useless, and only disseminated the evil over a larger area.'[173] Many attempts were made to control air pollution in the following years, but it wasn't until the 1950s that the issue became political on an international scale.

The Swedish scientist Oden first drew attention to the fact that acidification of lakes and rivers was connected to the long distance transport of sulphur dioxide and nitrogen oxides in the atmosphere. These two chemicals react to form sulphuric acid and nitric acid, and are deposited on soils, vegetation and in groundwater either wet (as rain, snow, fog or mist) or dry (by diffusion, or as acid aerosol particles such as ammonium sulphate and ammonium nitrate).

The biological decline of thousands of kilometres of waterways in Scandinavia had been well documented by the 1970s. Acidification, and the consequent mobilization of toxic metals such as aluminium and cadmium were taking their toll not just on fish and insect populations, but also on birds such as grebes and diving ducks. As host of the 1972 UN Conference on the Human Environment, Sweden grasped the opportunity to impress upon a sceptical world, the importance of the increasing acidity of European rain. Partly as a result of this, the OECD launched a major international research programme on the nature of cross-border air pollution in Europe. This in turn led to the signing by 35 countries of a revolutionary new treaty in November 1979.

The Geneva Convention on Long-range Transboundary Air Pollution (LRTAP), signed under the auspices of the UN Economic Commission for Europe (UN/ECE) was the first international air pollution agreement, and it provided a focus for environmental campaigns against acid rain in the following decade. The international research effort throughout the 1970s and early 1980s brought forth a welter of scientific evidence that aquatic ecosystems were being devastated and that populations of sensitive fish such as trout and salmon in Scandinavia, and burbot, chub and yellow perch in parts of North America were plunging. Nevertheless, action to reduce emissions of the main acid rain pollutants, SO_2 and NO_x was blocked by most of the industrial nations, including two of the biggest polluters, the UK and West Germany.

The Scandinavians tried again to break the political log-jam in 1982. On the tenth anniversary of the Stockholm Conference, they convened a Conference on Acidification. It was at this meeting that the West German Government performed its celebrated 'road to Damascus' conversion. Brought on by the documented decline of the German forests, and the associated rise of Green politics, this change of heart put West Germany firmly in the Scandinavian camp. Equally firmly, it lined up the

UK and the USA as the heavyweight opposition to pollution controls.[174]

Eventually, the development of an internal pressure group (the '30% Club') within the LRTAP Convention – originally including the Nordics, West Germany, the Netherlands, Austria, Switzerland and Canada – all of whom styled themselves 'victim states', forced the signing of a binding protocol to reduce sulphur emissions.[175] The Helsinki Protocol requires signatories to reduce sulphur emissions by 30 percent based on 1980 levels by 1993. Britain was the only major West European sulphur emitter to remain resolutely outside the protocol for the full term. A new and stronger sulphur protocol, aimed at controlling acid rain well into the next century, was agreed in March 1994.

Soon after the original protocol was signed the Swedish poet, Folke Isaksson, wrote of the still unsolved problem:[176]

> *this is an occupation, a silent war. It does not yield dark banner headlines, although it is a tyrannical develop-ment...if the forests die, the world first becomes inhospitable and then uninhabitable. And if the creations of the past are annihilated in this quiet storm assault, our guideposts will dissolve.*

It was at the 1985 signing of the first sulphur protocol in Helsinki, that British environmentalist Chris Rose first gave the sobriquet 'Dirty Man of Europe' to his country's government. It is a title still well-deserved and with currency. By 1990, UK sulphur emissions had only come down by 23 percent from 1980, and were on the rise again from the previous year.[177] There is a certain irony in the British situation. Acidity of rainfall was first recorded as a problem in England in 1662,[178] the term acid rain was coined there in 1872, and yet, 120 years later, the country is still Western Europe's biggest exporter of acid rain. In 1992, the UN reported that Britain's forests have the fastest rate of decline in Europe. The government itself admits that critical loads are exceeded in nearly 10 percent of Britain's freshwa-ters[179] and has admitted that a quarter of the most important conservation sites in England are threatened. What price now William Blake's 'green and pleasant land'?

A Climate of Change

The climate change debate came of age at the United Nations Conference on Environment and Development (UNCED) in June 1992, when 154 countries signed a treaty aimed at reducing the speed with which our planet warms. Coming as the culmination of more than a century of speculation, scientific debate and political negotiation, the UN Rio de Janeiro Framework Convention on Climate Change (UNFCCC) opened a new era of international cooperation to reduce atmospheric pollution.

Having received the requisite 50 ratifications, the new treaty formally entered into force on 21 March 1994, and signatories to the convention already find themselves mired in a debate that encompasses the central social and economic issues of our time. These include the global problems of poverty, population, and overconsumption of resources; policies on forestry, energy and transport; the politics of aid, trade, finance and technology transfer; all compounded by the uncomfortable political fact that the worst consequences of climate change will be felt by future generations and poorer countries, yet much of the pollution was caused in the past by rich countries.

But what is climate change? Frenchman Jean Fournier had identified the 'greenhouse effect' in the nineteenth century, and the Swede Svante Arrhenius called attention to the possible environmental impacts of doubling the level of carbon dioxide (CO_2) in the atmosphere as early as 1896. It was in the few years leading up to the UNCED Conference in Rio, however, that the phrase 'global warming' was on people's lips the world over. After the slow-burning issue of acid rain and the high pressure, panic years of dealing with the depletion of the stratospheric ozone layer, governments finally began to get to grips with global warming in 1988 when they established the Intergovernmental Panel on Climate Change (IPCC).

The IPCC was charged with providing a comprehensive review of the state of scientific knowledge on climate change. Two years later, after inputs from nearly 400 scientists in 25 countries, they reported that they were 'certain' that the greenhouse effect existed, that it was the mechanism by which the Earth remained warm enough for human life, and that pollution

70

by humankind will cause the world to get warmer. The IPCC report went on to say that immediate reductions of at least 60 percent in emissions of the main greenhouse gases – CO_2, nitrous oxide and chlorofluorocarbons (CFCs) – would be needed merely to stabilize their atmospheric concentrations at today's levels.[180] The predictions that emerged from the IPCC's sophisticated computer modelling exercises pointed towards a rate of temperature rise of 0.3°C per decade and sea-level rise of about 6 cm per decade.

The mechanism of global warming is sparked by the build-up of greenhouse gases in the upper atmosphere. This invisible blanket of gases allows short-wave radiation from the sun to pass through to the Earth's surface, but when a portion of the energy is reflected back towards space as long-wave infrared radiation, the blanket captures some and bounces it back to the land again. The more greenhouse gases build up in the atmosphere, the more heat fails to escape from the Earth, and the closer climate catastrophe comes.

A 1992 update of the IPCC report confirmed the earlier findings, highlighting the fact that there has already been a global temperature increase of approximately half a degree centigrade over the last century and that the late eighties and first two years of the nineties were the hottest years on record. In addition, the IPCC update recorded a loss in global snow cover of 8 percent over the last two decades.[181]

The potential consequences of global climate change are prodigious. Sea-level rise, even under conservative scenarios, could flood millions of hectares of coastal land destroying communities and obliterating agricultural land. Whole nations, like the Pacific states of Tuvalu and Kiribati, built on low-lying atolls, could be annihilated. Agriculture will be radically affected through the shifting of suitable crop climatic zones – sometimes by hundreds of kilometres or across borders and into other countries. One study suggests that the grain baskets of the US Great Plains and corn belt, the Canadian prairies and southern Russia could have yields depressed by up to 30 percent due to reduced soil moisture.[182]

Increased frequency and intensity of extreme weather events could lead to human tragedy and a growing stream of 'environmental refugees' in the aftermath of hurricanes, floods and droughts. Health impacts are also likely to be severe, with tropical diseases such as malaria and sleeping sickness able to expand their ranges.[183]

Climate Change and Biodiversity

Ecosystem damage may already be happening. Ocean warming (combined with other pollution stresses on the marine environment) is blamed by many scientists for a worldwide rash of coral bleaching , leading to the death of reefs. Most corals are already living in water close to their thermal tolerance levels, and warmer water can cause them to eject their symbiotic organisms and die.[184] Coastal mangrove forests may also be threatened. Crucial as fish spawning zones and important as sources of tannins, wood and honey, these tropical salt-water forests also provide protection against tidal waves for coastal communities in developing countries such as Bangladesh and India. Marine Biologist Joanne Ellison has estimated that many mangroves cannot survive sea-level rise of greater than 12 cm per century. The sea is expected to rise at five times that rate in the next few decades, and the first signs of mangrove decline have already been documented in Bermuda.[185]

Protected areas and national parks, many of which are already islands of biodiversity in a sea of urban and agricultural development may become useless as conservation mechanisms as the geographical ranges of the species they purport to protect shift out of the boundaries of the reserve, and as their characteristic ecosystems and species assemblages break up. In the process, years of conservation effort and expenditure will go to waste. Even the most conservative analyses show that more than half of the world's 'biosphere reserves' for instance, would be significantly changed by the greenhouse effect. Alpine flower meadows, Arctic marine ecosystems, tundra, tropical cloud forest, and coastal saltmarsh have all been classified as highly vulnerable to climate change.[186]

Of these systems, perhaps the least attention has been given to the effects that climate change may have on tropical forests, and in particular, tropical cloud forests. Little is known at all about cloud forests as they are generally found perched on the peaks and ridges of the most remote mountain ranges. But as deforestation of the lowlands proceeds apace, conservation attention is focusing more on these high mountain retreats. Animal and bird species from lower down the slopes are being forced to move uphill and seek refuge in the unfamiliar forests of the high montane zone. These forests have a character all their own, since the cloud that envelops them most of the time saturates the air and soil, and the trees that can eke out an existence in this hostile environment are often dwarves that form a canopy only a few feet off the ground. To walk through a tropical cloud forest is to enter a cool, quiet world of drenched hanging mosses, and a profusion of orchids and bromeliad ferns perched on trunks and branches. A footprint left in the soggy mat that is the forest floor can last for twenty years, and full recovery from a hurricane hit could take more than two hundred years.

Climate change poses severe threats to these most sensitive of forests. The most obvious comes from local warming. If temperatures rise, then so too could the cloud caps that cover mountains such as Kinabalu in Malaysia or Luquillo in Puerto Rico. Without cloud, there can be no cloud forest. Studies in the remote Xishuangbanna forests of China's southern Yunnan province have shown that increased sun and reduced fog and cloud during the last 15 years are already causing the loss of some orchids and other plant species.[187]

A second threat comes to forests that fall within the tropical cyclone belt that runs around the equator. The warming of the sea surface that comes as a result of the greenhouse effect is expected to increase the intensity, size and frequency of tropical storms and cyclones. Hurricanes may also occur further away from the equator and last longer, penetrating forests where they have never hit before, and where, as a consequence, the tree species have not evolved to withstand them. Hurricanes can devastate tropical forests. The Kolombangara forest of the Solomon Islands lost almost all its large trees in 1967. Hurricanes Gilbert and Joan in 1978 and Hurricane Hugo in 1989 flattened huge expanses of the forests of Jamaica, Nicaragua and Puerto Rico respectively.[188]

Resilience to hurricanes is remarkable, and most tropical

forests below the cloud zone can recover fast, even from the most severe storm damage. The problem comes when the winds hit more frequently than normal. Some studies suggest that if hurricane frequency doubles, then tree deaths will increase by 50 percent. Where the storms don't just come more often, but also rage more ferociously, the toll could be much higher.[189]

Tropical forests offer just one example of the habitats that can be listed in a catalogue of potential disaster and devastation. But from the growing knowledge and concern about these threats to our natural heritage was born the UN Framework Convention on Climate Change. Highly ambiguous and open to all sorts of interpretations, the treaty does nevertheless have an extraordinarily clear objective:[190]

> ...*stabilization of greenhouse gas concentrations in the atmosphere at a level that would prevent dangerous anthropogenic interference with the climate system. Such a level should be achieved within a time-frame sufficient to allow ecosystems to adapt naturally to climate change...*

A scientific debate is already raging as to how drastic are the measures that will be required to meet this objective. Under the original 1992 commitments for the convention, developed nations would be required to stabilise their greenhouse gas emissions at 1990 levels by the year 2000. At negotiating sessions in Geneva in February 1994, industrialized countries acknowledged for the first time that these existing commitments 'should be considered inadequate in the long term'.[191] Countries including Denmark and Germany proposed that a target of 20 percent reductions should be set for the year 2005, but many other nations, whilst prepared to admit that more stringent measures should be taken, have shown themselves reluctant to set targets for fear of not being able to implement effective policy measures to achieve reductions.

In the USA, for instance, the publication of President Clinton's *Climate Change Action Plan* in October 1993 was greeted with very muted enthusiasm by environmentalists. The plan, although more detailed than those of other major greenhouse gas emitters such as Japan and Germany, relied mainly on voluntary action by industry, and almost completely failed to address the critical issue of reducing car exhaust emissions. The report authors admitted that 'It relies on the ingenuity, creativity and sense of responsibility of the American People.'[192] In the eyes of some, this provided a sure sign that the plan would be destined to fail in a country where most working people rate the fear of unemployment much higher amongst their concerns than the environment.

As politicians struggle with the political unpalatability of regulating greenhouse gas emissions, the IPCC has launched an international scientific investigation to try and determine what the critical thresholds of climate change are for various ecosystems. Some assessment of just how versatile plants and animals will be in adapting to global warming will be essential if governments are to start putting long-term environmental benefit before short-term electoral gain.

Radioactivity: the Unseen Killer

Ever since Enola Gay laid her lethal egg on Hiroshima in August 1945, atomic power has been in the news. Sixty-four thousand Hiroshima residents died within four months of the bomb going off, a third of them from radiation sickness. The tragedies of Hiroshima and Nagasaki provided the nadir to the exhilarating years of nuclear discovery in the 1930s and 40s.

The theory of nuclear fission had been developed by the Germans, Strassman and Hahn in 1938. They proved that the nucleus of the uranium atom splits, releasing huge quantities of energy when bombarded with neutrons. Not long afterwards, Frederic Joliot-Curie showed that fission could set up a chain reaction of self-sustaining atomic explosions and other scientists contributed the information that a minimum amount, or 'critical mass', of uranium was needed for fission to take place.[193]

As war drew closer, thoughts of nuclear bombs became more real, and the refugee scientists Otto Frisch and Rudolph Peierls confirmed that a bomb weighing only 5 kilos and containing the rare isotope uranium 235 could be made and would work. But, they wrote, 'the whole material of the bomb would be transformed into a highly radioactive state... and the radiations would be fatal to living beings even a long time after the explosion.'[194]

While research raced ahead in both Britain and Germany, the then president of the Royal Society wrote to Churchill 'I cannot avoid the conviction that science is approaching the realisation of a project which may bring either disaster or benefit on a scale hitherto unimaginable to the future of mankind.'[195] This was a prescient view of where the bomb took society after the war. The uncertainty about the beneficial as opposed to the harmful impacts of nuclear power form the basis for much of today's energy debate, and the proliferation of nuclear weapons remains a significant threat to this day.

The harmful effects of atomic, or ionizing radiation, have been known since the early discoveries of Roentgen, Becquerel and Madame Curie, but the precise impacts and mechanisms of nuclear pollution are still not fully understood. Radiation penetrates biological matter and acts on the cells and their consituent parts by causing chemical, molecular or physical damage often resulting in cell death or genetic mutation. Unlike most toxic chemicals, with radiation there appears to be no level of dose below which damage cannot be caused.[196]

The fear of both nuclear war and radioactive contamination was the launching force behind several citizens' groups in the 1950s and 60s. The Campaign for Nuclear Disarmament (CND) marched against Aldermaston with Michael Foot at their head, and Greenpeace's first actions involved sailing its boats against the nuclear tests at Amchitka and Mururoa. The United States exploded bombs at Bikini Atoll in 1946, having relocated the island's entire population, and the British began testing in Australia's Monte Bello Islands in 1951. Between that date and 1963, the US exploded 93 test bombs in the Nevada desert.

Until the early 1960s governments' awareness of the potential long-term impacts of radiation pollution seemed almost nil. Even after 1963 when the limited test ban treaty banished nuclear testing underground, France and China continued to ignore the agreement and test in the open atmosphere. In his book *No Conceivable Injury*, Robert Milliken quotes as conservative, estimates by Robert Alexander of the US Nuclear Regulatory Commission that the combined effects of the testing programmes have been nearly 85,000 cancer deaths and 168,000 genetic defects. It wasn't until the mid-1980s that victims of cancers resulting from proximity to testing began to be recognized officially, and some compensation awards made.[197]

Since the 1960s, the campaign against the use of civil nuclear power has been one of the uniting forces in the environmental movement. While the nuclear industry has claimed that nuclear power is one of the safest and most efficient ways of producing energy, environmentalists have countered with arguments about the impractibility of disposing of waste that remains dangerous for millenia, and the unique pollution threat caused by accidents at nuclear power stations.

Even a World Health Organization panel on which the nuclear lobby was heavily represented, recently reported 'appropriate methods for the final disposal of high-level radioactive waste have yet to be implemented'. And it isn't just waste from nuclear power stations that is causing the problem. The careless disposal of a gamma-ray radiotherapy machine in the small Brazilian town of Goiania in 1987, and its subsequent dismantling by a duo of opportunistic scrap metal dealers, led to the death of four people, irradiation of a further 250, evacuation of 41 homes and the creation of 275 truck loads of nuclear waste.[198]

After 40 years of production of nuclear waste, it is little wonder that there is 'considerable public anxiety' over nuclear power. Of course, the near melt-down at Three Mile Island, slow pollution of the North Sea by Sellafield (ie Windscale) and the terrifying explosion at Chernobyl haven't done much to lessen public anxiety. 1992 alone saw reports confirming radioactivity from Sellafield to be a major factor in the abnormally high child leukaemia rate in the local village of Seascale, and also showing 'unexpectedly high' levels of thyroid cancer in Russian and Ukrainian children caught in the fallout from Chernobyl.

Chernobyl caused the greatest industrial release of radioactive pollution ever, spreading heavy fallout for more than 1500 km. A low estimate for consequent fatal cancers is 28,000. Seven hundred million people live within 160 km of one of the world's 400 nuclear power stations. Still the Council for Energy Awareness can blithely advertise: 'Every day is Earth Day with nuclear energy'.[199]

Chapter Seven

AN ALL CONSUMING PASSION

*We who have so much, to you who have so little, to
you who don't have anything at all.
We who have so much more than any one man does
need, and you who don't have anything at all.*

Lou Reed

The Consumer Class

In the lead up to the 1992 US Presidential Election, Gore Vidal
wrote that the incumbent, George Bush, saw himself in the
'heroic line of the great and the good and the right. But the world
of the 90s does not resemble, in any way, the 30s. There is no
Hitler, no Stalin. There are no regnant ideologies other than our
own, which is consumerism.'[200]

Post-war America is the undisputed champion of material
consumption. Where the US consumer leads, the rest of the
world aspires to follow. Bill Clinton presides over an America
that eats more meat, uses more water, drives more cars, watches
more TV and produces a greater volume of domestic refuse than
any other civilization has ever done.

In the shadow of the United States come the other OECD
countries, all of them in the top league of global consumers. The
consumption patterns of the rich west would be completely
unsustainable if extended to the rest of the world's population,
and yet per capita consumption is steadily growing in virtually
every country in the world, and the affluent in many developing

countries already have the spending power to put them on a par with the best of the West.

Worldwatch Institute's Alan Durning identified three world consumption classes in his book *How Much is Enough?*[201] First come the poor. Numbering 1.1 billion, they earn 2 percent of the world's income, eat almost no meat, drink dirty water, travel mostly by foot and struggle to find adequate shelter. Then comes the middle-income class, who account for a third of global income and include 60 percent of humanity. They have an adequate diet, better sanitation, solid housing, access to electricity and basic stocks of durable consumer items. Durning's third group is the consumer class, and they are us: the 20 percent of the world's population which earns 64 percent of its income, lives on meat and processed food, which drives cars and flies off for vacation taking along the latest in consumer gadgetry.[202] The consumer class is also responsible for much of the world's water pollution, most of its toxic and domestic wastes, for the manufacture of the vast majority of synthetic chemicals, and is the source of approximately 80 percent of global air pollution. Pollutants are now the chief products of the consumer society.

Nature, believed Thomas Paine, had made man's wants 'greater than his individual powers. No man is capable, without the aid of society, of supplying his own wants.'[203] But just exactly what *do* we want, and what are we prepared to sacrifice to get it? Society has concentrated during the last half century on providing the basis for material consumption. Most people's aspirations now revolve around increased acquisition of possessions and growing individual consumption. Yet while everyone would agree there is a certain 'comfort level' below which human communities should not have to dip, there is no evidence that the crazy rush to consume and possess has brought happiness to those that have pursued it.

One psychological study showed that Americans are no more happy than Cubans, but both are happier than Filipinos, West Germans and Indians. Satisfaction with life tends to be more related to where people stand in their own society, than to their overall level of income or consumption in the global context. 'Rising prosperity in the United States since 1957 has been accompanied by a falling level of satisfaction' says psychologist Martin Argyle, and 'international differences in happiness are very small and are almost unrelated to economic prosperity.'[204] Comparing the societies of East and West Germany almost a decade before the Berlin wall fell, Rudolf Bahro commented that 'more important than the quality or quantity of consumer goods...is the need for a new consumption pattern geared to the qualitative development of the individual.'[205]

Consumption acts like an addictive drug on our societies. The consumer class drives itself towards greater highs, but the cravings become increasingly hard to satisfy, the empty feeling between fixes continues, and the consumer never, ever, gets enough. Shopping is the apotheosis of the consumer culture. And, to stretch the narcotic analogy further, retailers and whole-salers become the dealers to a hooked populace while in the hazy subculture of the marketing men, new desires are created and catered for.

This modern consumer malaise grew from America's struggle to climb from the depression and the dustbowl up J K Galbraith's 'mountain of well-being.'[206] As Americans had more to lose, the focus of economic theory became security, and with this tenet came the primacy of production. If production could be maintained, irrespective of actual need for products or services, so too could employment, and in order to maintain production, desires had to be created.

This economic revolution spawned a situation where even in 1992, President Bush could prepare for the Earth Summit – the largest gathering of heads of state the world has ever seen – by claiming as he got on the plane to Rio, that jobs came before environmental protection. His Vice President, Dan Quayle, headed up the Competitiveness Council, a White House unit set up specifically to block and dismantle environmental legislation perceived by business to be hindering it in the pursuit of produc-tion. All this followed a decade from 1975 to 1985 when more than 300 of the 'Fortune 500' companies had been convicted of serious environmental pollution offences.[207]

After UNCED, President Bush was one of the proud progenitors of the North American Free Trade Agreement (NAFTA) which allowed US companies to decamp to Mexico and maximize profitability by avoiding US pollution laws and contaminating the land and water to the south of the Rio Grande. This wasn't quite what Galbraith meant when he said 'the importance of production transcends our boundaries'. He went on to say that Americans are continually told that their 'standard of living is the marvel of the world'.

It may not be the marvel, but it is certainly the dream of millions. The advent of video and modern communications systems enable images of modern America to be seen in every corner of the world. Images which have often been carefully honed to take advantage of a new generation of consumers, a new market, a new life-support system for the secular god, production.

US citizens are yearly bombarded with $131 billion worth of advertising on TV and radio, in newspapers and magazines, with their cornflakes splashed across billboards and dragged through the skies behind biplanes. J K Galbraith again:

> Were it so that a man on arising each morning was assailed by demons which instilled in him a passion sometimes for silk shirts, sometimes for kitchenware, sometimes for chamber pots, and sometimes for orange squash, there would be every reason to applaud the effort to find the goods, however odd, that quenched his flame' but 'production only fills a void that it has itself created.'[208]

The consumer boom has not been confined to the US. All the western countries have played their part in the movement. Every one of the 343 million inhabitants of the EC spends approximately $11,000 a year on consumption. Each EC citizen eats 94 kilos of meat, 116 kilos of cereals, 43 kilos of eggs, butter, cheese, cream and yoghurt, 130 kilos of bakery products, 43 kilos of canned and frozen foods and much, much more. They drink 74 litres of milk, 260 litres of alcohol, fruit juices, bottled waters and canned drinks. They smoke 1700 cigarettes and 20 cigars annually while using 23 kilos of soaps, detergents and cleaning fluids.

Europeans buy more than one and a half billion pairs of shoes each year, 19 million fridges and freezers, more than 7 million cookers and as many microwave ovens. These go along with annual sales of 21 million colour TVs, nearly 11 million video recorders, 14 million cameras, 4 million electric blankets, 16 million hair dryers and 6 million toasters.[209] History does not yet record how many Mickey Mouse peanut dispensers, dancing Coke cans, plush Garfield car mascots or electronic dog whistles the average European buys.

The modern consumer seems to satisfy a bulimic urge to gorge on goods, vomiting forth a constant stream of waste packaging, obsolescent durables and rubbish, before returning once again with an empty feeling to binge at the trough of commercialism.

The amount of garbage flowing from the spending frenzy is prodigious – approximately 350 kilos annually per capita in Europe – and it has been estimated that five times as much waste is generated in manufacture and 20 times as much at the site of original resource extraction.[210] Apart from the pollution, the resource wastage is phenomenal. Domestic water needs, for instance, could be met almost everywhere in the world with about 100 litres per person per day. It takes this much water to produce a kilogram of paper, and 50 tons of water to make a ton of leather. Yuppie environmentalists in their Barneys or Boss suits should take note that worsted wool is 50 times as expensive in terms of water as even leather.[211]

The rate of increase in per capita consumption has been staggering during the last five decades, but signs that things were changing can be detected in both American and British fiction of the period. In *Coming up for Air*, George Orwell's insurance salesman of 1939 goes into a milkbar and accidentally orders an ersatz frankfurter filled with fish paste:

> *It gave me the feeling that I'd bitten into the modern world and discovered what it was really made of. That's the way we're going nowadays. Everything slick and streamlined, everything made out of something else. Celluloid, rubber, chromium steel everywhere, arc-lamps blazing all night, glass roofs over your head, radios all playing the same*

tune, no vegetation left, everything cemented over, mockturtles grazing under the neutral fruit-trees.'[212]

This growing feeling among Orwell's generation, one of being steadily alienated from nature and from increasingly artificial surroundings, was reflected also in John Steinbeck's masterpiece *The Grapes of Wrath.* Describing the industrialization of agriculture, he talked of the hired tractor drivers and the execution of their job:

> *not plowing but surgery.... The driver sat in his iron seat and he was proud of the straight lines he did not will, proud of the tractor he did not own or love, proud of the power he could not control.... Men ate what they had not raised, had no connection with the bread. The land bore under iron, and under iron gradually died.[213]*

As the great production drive got under way, the transformation of America's corn-belt agriculture into the mechanized monster of the late twentieth century began. The dustbowl was born, and millions of tons of topsoil blew from the land creating the duel problems of loss of productive agriculture and dust pollution, the latter radically increasing respiratory illnesses. Even today, soil erosion is one of the most serious problems facing US agriculture.

The separation of people from the means of production has been a key factor in transforming western nations into ones populated by captive consumers. Dietary habits in the developed world use more energy, exhaust more land and create more pollution than ever before. They have also brought with them modern diseases such as coronary heart disease, tooth decay and various cancers.

The Death of Food

Convenience food has become the mainstay of the modern diet, and without exception it creates more pollution than the more common, locally grown fresh food of pre-war years. A survey undertaken by *Harpers* in April 1992 revealed that 74 percent of Americans don't know how long it takes to hard boil an egg.

Microwave food may not take much energy to prepare at home, but it more than compensates in its production and packaging. Canned food uses large quantities of water and energy at the factory, and even when fresh food is available it burns fossil fuels in its transport from the farm gate to the family table. According to the Worldwatch Institute, three times as much energy is used transporting a lettuce from California to New York, as goes into growing the vegetable in the first place. Typically, says Worldwatch, food in the US travels 2000 km before it is eaten.[214]

The products of McDonalds and Coca Cola are the archetypal convenience foods. Aggressively marketed throughout the world, they spearhead the consumer revolution wherever they arrive, and provide a far from subliminal advertisement for a particular lifestyle. At least in the United States, McDonalds has recently made significant efforts to reduce its waste output. It even launched a campaign titled 'McRecycle USA' in 1990.[215]

The hamburger business is important to a discussion of pollution and consumption for several reasons. The first of these being that the hamburger society has been a major force behind the massive growth in the amount of meat eaten in western countries. Alan Durning records that every kilogram of beef produced requires 5 kgs of corn and soybean meal, 3000 litres of water and two litres of gasoline.[216] A huge amount of fertilizer is also used in the production of the feedstock and on pasture.

The other crucial impacts of the ground beef business have been in the field of technical and business innovation. In his truly revolting essay 'Anatomy of a Cheeseburger', Jeremy Rifkind credits the Chicago meat-packers as being the first industry to use modern assembly lines, thereby pioneering the technology need to fuel the consumer boom a few decades later. He also records how Upton Sinclair's 1904 novel exposing the unsanitary and polluted conditions of the Chicago slaughterhouses, resulted in new national legislation aimed at preventing the sale of contaminated meat within two years of the book's publication.[217]

The acknowledged stroke of genius of McDonalds, however, was not the standardization of the 1.6 oz, 3.875 inch diameter premiere product, nor even the restricted choice of menu items;

it was the fact that people could take their food away in paper bags. The ultimate marriage of convenience between the car and takeaway food packaging allowed the stellar growth of the fast-food industry from the 1950s. Today, more than 17 million hamburgers are sold every day in the US alone. There is a US McDonalds outlet for every 29,000 Americans, and somewhere in the world the company opens a new store every 18 hours, yet James Cantalupo, president of McDonalds International can still say 'I don't see any market where we can't continue to accelerate our growth.'[218]

Along with the hamburgers go the soft drinks. Per capita consumption in the US is around 80 litres a year. This is the same as Australia, twice as much as Brazil, nearly twenty times more than in Zimbabwe and 100 times as much as in Indonesia.[219] Worldwide consumption rose by nearly a third between 1980 and 1990. Even 'fresh' orange juice is not so innocent. The developed world drinks 90 percent of the world's orange juice, and 80 percent of that drunk in the highest consuming nation, western Germany, comes from Brazil. Apart from the energy used to transport the juice 12,000 km, there is also a high resource cost to the concentration process. Each litre of Brazilian orange juice drunk requires 22 litres of water to produce. Every litre of US juice (from Florida) needs up to 1000 litres of irrigation water and 2 litres of fuel.[220]

Just one drinks business, The Coca Cola Company, controls 45 percent of world carbonated drinks sales. Coke was invented by Atlanta patent medicine quack and morphine addict, John Pemberton, in the late 1880s, and went on to become the world's most popular soft drink. Along the way, Coke dropped the stimulants cocaine and kola nuts, and became a formative force in the development of modern advertising and marketing. Just like tea, sugar and tobacco before it, Coca Cola was originally sold as a pick-me-up and medicine. Transformation over a century has made it 99 percent sugar-water with caramel colouring, and the centre of an annual 4 billion US dollar marketing campaign. The history of Coca Cola has paralleled the Second Consumer Revolution, and taken the product to the cutting edge of what will be the next phase – the globalization of US-orientated culture and consumption patterns.[221]

Packaging the Dream

The problem with these drinks is not so much the contents as the 200 billion containers thrown away every year. Many of these discarded containers are made of aluminium. The aluminium can first came on the scene in 1963 and now fully one quarter of US aluminium supplies are used for drinks packaging. Of these about half end up in land-fill waste sites. Unfortunately, aluminium uses more energy to produce than any other metal. The aluminium produced for every six-pack of beer uses electricity equivalent to one beer can full of petrol, and the world aluminium industry uses as much electricity as the whole continent of Africa every year.[222] Bauxite, from which the metal is extracted, was first discovered in the picturesque French hilltop village of Les Baux en Provence. While Les Baux is now a thriving tourist destination, the hunger for aluminium has devastated the environment of many developing countries.

Production of aluminium shot up nearly fourfold in the three decades after the development of the disposable can, and despite the fact that enough is produced to provide three kilograms every year for every person on earth, the industrialized world still uses 86 percent of the total and maintains a per capita consumption rate 19 times that of the developing countries.[223] The need for energy to smelt aluminium has been behind the development of many major dam schemes, including the disastrous Akosombo Dam in Ghana, Guri in Venezuela, the infamous Tucurui Dam and Grand Carajas development in the heart of the Brazilian Amazon, and the James Bay hydroelectric project in the Cree Nation lands of Quebec.

Aside from aluminium cans, packaging of all kinds is now under the spotlight of environmentalists' attention. In the US, packaging waste accounts for roughly one third of municipal solid waste, while in the Netherlands domestic packaging is thought to make up 22 percent of the total waste stream. It is extraordinarily difficult, however, to get accurate statistics on the world's rubbish. One US 'garbologist', William Rathje, commented 'we have more reliable information about Neptune than we do about this country's solid waste stream'.[224]

The fact that packaging pollution is moving up the political agenda in Europe also can be seen from the new (1991) German Packaging Recycling Ordinance, which requires that manufacturers must take back and recycle packaging materials. Targets have been set for the compulsory recycling of between 64 percent and 72 percent of all packaging materials by June 1995. Less than a year after the law came into effect, there had been a significant move from the use of polyvinyl chloride (PVC) and other plastics to paper and card, and a trend towards reduced packaging overall.[225]

For the Netherlands, Friends of the Earth report that national packaging sales are growing by 4 percent annually, and that the greatest sectoral increase is in plastics, with 11 percent growth. The manufacture of Dutch packaging uses about 2 percent of industrial energy, is responsible for 5–10 percent of air pollution emissions and causes 2.5 billion cubic metres of water to be polluted. In the plastics sector one of the most serious pollution problems is the use of pigments (based on toxic metals such as cadmium, mercury, chrome, zinc and cobalt) which are released into the atmosphere or incorporated in ash and solid residues when plastic wastes are incinerated.[226]

Of all the plastic packaging materials, PVC is the most dangerous. Roughly half the content of PVC is chlorine, and many chlorine production processes involve the use of mercury batteries for electrolysis, and consequent mercury pollution of air and water. The manufacture of PVC also leads to emissions of mutagenic dichloroethane, vinyl chloride and toxic plasticizers such as pthallates. Incineration of waste products containing chlorine can allow pollution by dioxin and furans, and yet PVC only offers very poor recycling potential.

Carrying out life-cycle assessments for various types of drinks packaging it becomes clear that the least polluting forms of container are returnable PET and glass bottles, with PET being the preferable long-term option. Recycling is better than outright wastage, but the simple truth is that re-use is best.

And what do we put in our plastic bottles? Well increasingly, we fill them with good old water. While more than 1.2 billion people lack access to safe drinking water, the consumer class of Europe glugs and gurgles its way through more than 17 billion litres of Perrier, Evian, Buxton, San Pellegrino and the rest every year.[227] The French, Italians and Germans drink more bottled water than anyone else, and in Britain (where it costs 1000 times more than tap water), sales trebled between 1988 and 1990.[228] US consumption increased by 16 percent in 1991[229] and the Japanese slurped almost twice as much in 1989 than they had the year before.[230] One in six American households now buys bottled water regularly, nearly a three-fold increase in a decade. Fear of pollution seems to be a driving force in the US market, and people are growing increasingly distrustful of their tap water.[231]

Trends towards healthier living, fashion and fear of polluted tap water are driving this revolution in drinking patterns, undoubtedly spurred on by advertising. Perrier alone spent $15 million in the UK market in 1990. As hundreds of thousands of people in Africa daily walk miles to the nearest source of water, jet cargo planes speed overhead burning avaiation fuel and taking Perrier and Evian to the thirsty masses of New Zealand, Japan and Singapore. Even in Indonesia the demand for bottled water is growing, and consumer groups have urged the government to prevent its suppliers exploiting a monopoly as Jakarta's groundwater supply becomes increasingly polluted and less drinkable.[232]

Plastics pervade the lives of the post-war generation and of the consumer class. The US has 700 times more plastic flamingoes in its backyards than real ones in natural habitats. In Japan individual fruits are packed on polystyrene trays and wrapped in clingfilm. Fresh orange and lemon juice in Switzerland comes in plastic bottles. No trip to the shopping mall or market is complete without a plastic bag. Electrical goods like video cameras, radios and CD players are encased in plastic. Carpets, clothing, greenhouses, furniture, lighting fixtures, luggage: all can be plastic and often are.

Use of plastic in the developed countries is 2000 percent higher than it was in 1950 and to achieve these Olympian heights of increased consumption, even vegetation has been plasticized. Astroturf has replaced grass in soccer stadiums, plastic tulips now vie with real ones for tourist attention in Amsterdam's flower market, and plastic palm trees adorn low-maintenance homes and hotel foyers.

Plastic waste is the consumer's epitaph. Many is the polystyrene hamburger clam-shell, supermarket shopping bag or Nintendo joystick that will last longer than the inscriptions on the tombs of Pere Lachaise or Highgate cemeteries. Plastics litter the world's coastline; bags and pellets float in the deep oceans. No roadside verge in the western world is free of plastic, and sometimes it seems no delineating barbed wire fence exists, which doesn't have its complement of ragged, grubby remnants of plastic bags and sheeting, skewered like gamekeepers' trophies and blowing brightly in the wind.

This plastic, this flexible and convenient material, which can last for centuries in the environment, is increasingly being found in that other icon of convenience, the car. A tenth of all plastics in the OECD are used for car parts – from seat, to dash, to oil tank – and this percentage looks set to increase.[233] The growth in the world's car population, combined with a trend towards lightness (for fuel efficiency) and eventual recyclability is making plastic the material of choice for car manufacturers.

But car makers will have to go a long way before they can claim their vehicles are environmentally benign. Apart from the air pollution threat from vehicle exhausts, there remains the problem of disposal of parts and old cars. Greenpeace estimate that 40 million cars are dumped every year, and 270 million new tyres produced. This implies the disposal of 240 million litres of lubricating oil, 25 million tons of steel, 5.5 million tons of aluminium, 4.2 million tons of polymers, 3.6 million tons of iron, 2.3 million tons of rubber in addition to assorted copper, brass, zinc, lead, rhodium, asbestos, plastics, solvents, glass and chlorofluorocarbons. More than 100 million car batteries are also disposed of every year, and they represent a special threat with their leaky combination of lead and sulphuric acid.[234]

J K Galbraith summed up our ambivalance to the car when he wrote 'who can say for sure that the deprivation which afflicts him with hunger is more painful than the deprivation which afflicts him with the envy of his neighbour's new car?'[235]

The Dark Side of Suburbia

The growth of the suburbs in the twentieth century allowed

consumers to live the dreams that were sold them by the forces of marketing. The fast-talking 'tin men', for instance, convinced half of suburban America to upgrade from wooden clapboard to aluminium siding during the 1960s. Outside their expensively converted houses, stocked with the latest in consumer durables, they could park the centrepiece of their lifestyle, the family car.

In 1920 most Americans lived in rural areas but 70 years later less than a quarter of the population did. The nation had matured through urbanization to suburbanization, made the choice of private space over public space and converted from front porches to back yards.[236]

Stephen Knight describes a similar process in Australia:[237]

a culture of desperate impoverishment had generated wealth of a grotesque character. Filling the swimming pool to build a tennis court, buying a four-wheel drive to fabricate a rural self, drinking imported mineral water for the healthy travelled self...a special symbol of that surplus society is the hapless third vehicle that has to squat in the gutter, because there are two others of its kind in the garage.

The homes of suburbia are filled with the fruits of consumer foraging. Whether seeking out preferred prey in the specialist stores or grazing in the malls, today's suburbanite lives to shop. Apart from watching TV, Americans devote more time to shopping than to any other activity, and much of the TV they see is advertising. Before they graduate from high school, US teenagers have watched over 100,000 adverts.[238]

According to Witold Rybczynski, most people choose consumption over free time. Put simply, they would rather work longer hours to buy things, than spend the time freely but without increasing their material possessions. In 1989, Americans worked through more than a billion hours of potential leisure time in order to spend $13 billion on 'increasingly elaborate running shoes, certified hiking shorts, and monogrammed warm-up suits'.[239] The 35,000 shopping malls where most of the money changes hands have been described by William Kowinski as 'the culmination of all the American dreams, both decent and demented, the fulfilment, the model of the post-war paradise'.[240]

But even the dream homes of the consumer class are built on the poisoned wastes of the industry that allowed suburban affluence to develop in the first place. The tragedy of Love Canal opened a new chapter in the pollution debate in the 1970s, when high rates of cancer and genetic disorders were found in residents of a housing estate built on the site of an old chemicals dump. Most OECD governments didn't even begin to regulate the disposal of toxic wastes until the middle of the 1970s, and by then billions of tons had already been dumped in hundreds of thousands of sites throughout the world.

The legacy of this cavalier attitude to disposing of the toxic residues of the consumer society is frightening to contemplate. Love Canal has been followed by a string of similar cases in other nations. In 1980, 900 people had to be permanently evacuated from their homes in Lekkerkerk, near Rotterdam, when the soils and groundwater under their homes was revealed to be contaminated by toluene, xylene and heavy metals. Residents of Hamburg were dismayed in 1979 to find themselves atop enough abandoned nerve gas to kill the entire population of the city. Two years later, hundreds of tons of carcinogenic organic solvents – wastes from a paint and varnish factory – were found buried in the centre of Copenhagen.

According to the World Resources Institute in Hebden Bridge in Yorkshire, at least 70 people have died as a result of diseases related to the unrecorded dumping of asbestos wastes from a textiles plant. Similarly, children of workers from the Blue Sky Mine in Australia used to play in piles of deadly blue asbestos waste before their parents' homes were bought up and bulldozed so that no one could again live in them after the mine closed.

Chapter Eight

ENERGY AND SURVIVAL

The Germans are dying out.
Living space without people.
Is such a thought possible?

Günter Grass

Clean Energy?

At the very core of our current environmental crisis is an energy dilemma. How do we utilize our energy resources to achieve development and provide improved life quality, without causing long term or even irreversible environmental damage? Some of the most intractable pollution problems have at their root the untrammelled exploitation of fossil fuels. Acid rain and climate change, urban air quality and photochemical smog are all primarily caused by the burning of coal, oil, or gas.

Non-fossil sources have their impacts too. The accident at Chernobyl in 1986 which spread a giant radioactive cloud over Europe is just one example of the dangers of nuclear power. Close to melt-down, the reactor spewed enough radioactivity into the atmosphere to turn the surrounding Ukrainian country-side into a no-go area, populated only by mutated flora and fauna. As far away as northern Scandinavia, the native Lapps were banned indefinitely from harvesting their reindeer herds, their prime source of meat. The latest studies from the Ukraine and from the World Health Organization (WHO) concur that already a massive, and unexpected, increase in thyroid cancers is being manifested among children in Russia who were exposed to radioactive iodine in the few days after the accident.

Even hydroelectric power, often cited as a form of 'clean' renewable energy, has its problems. These don't all lie with pollution: for instance, dams often hold back the silt that flows down rivers to renew the fertile plains and estuaries downstream. In *The Dammed*, Fred Pearce recounts how the building of the Akosombo Dam in Ghana, apart from being an economic disaster for a government in the grip of a US aluminium corporation, brought ruin to people on the coast. Where once the waters of the Volta had provided the silt that formed the coastal strip, the currents of the Atlantic carried away the land. The 10,000 strong population of Keta saw their sand-spit town disappear in a matter of years.[241]

Similar stories can be told about dams the world over. In terms of pollution, they can be said to have two major impacts. Firstly, the flooding of the land behind the dam should be considered as pollution. The 10,000 megawatt Guri Dam in Venezuela, for instance, flooded hundreds of square kilometres of tropical forest, and the dams planned for the Danube would destroy some of Europe's last untamed riverine flood forest. The proposals for the Three Gorges Dam in China and the Narmada Dams in India would lead to millions of people losing their homes and livelihoods as their land is 'polluted' with water and they undergo forced removal.

Second, in a more traditional pollution context, the disease which can develop as a result of the build-up of stagnant water behind tropical dams is significant. WHO estimates that more than 200 million people are affected by the debilitating and often fatal snail-borne disease schistosomiasis, and that the disease's endemic spread 'has been severely influenced by water construction works'. Even the building of the Aswan High Dam in 1900 increased schistosomiasis infection by an order of magnitude in the local Egyptian population.[242]

Geothermal energy, too, has been promoted as a less environmentally damaging form of energy production, and is an increasingly attractive option in countries where the natural heat of the earth can be tapped. Beta Balagot, in a comprehensive review of the impacts of Philippine geothermal energy projects, identified noise pollution as a potentially major health hazard, air pollution by hydrogen sulphide vented with geothermal steam, and dangerously high levels of arsenic and boron in waste waters. The question of siting of geothermal plants has also been raised to the top of the political agenda in the Philippines (the world's second largest producer of geothermal energy) with the proposal to build a plant in the sacred tribal land and tropical forest of Mt Apo.[243]

Not even the true renewable energy sources such as wind, waves and solar escape criticism. Current technology proposed for coastal wave power in the UK would involve mile upon mile of bobbing generators the size of two-storey terraced houses, just off shore. And potential wind and solar farms are consistently opposed by locals in the proposed development areas on the basis of visual intrusion and loss of amenity.

A History of Excess

From the sixth century AD, the use of human energy (often slave-power) and animals such as horses and oxen began to be replaced by water power. By 1068 there were more than 20,000 water mills in France and nearly 6000 in England. Towards the end of the twelfth century, water power was being supplemented by wind power and windmills first appeared near the coasts of the North Sea and the Mediterranean.[244] The various types of mill were used chiefly for grinding wheat and for tanning and fulling, and then increasingly for operating blast furnace bellows. A growing demand for thermal energy, required for the manufacture of iron and the new industries based around glass and tile-making, contributed to rapid deforestation in the Europe of the middle ages as timber was cut for fuel. In terms of pollution, the worst that woodfuel caused was a local smoke problem.

Major change in the European energy system came as demand for woodfuel began to outstrip supply. In London, for example, coal was available, but was not a popular fuel in the early thirteenth century. Lime burning for the construction industry was one the chief consumers of firewood, traditionally oak-brushwood. Population growth and conversion of forest land to agriculture after the Norman invasion probably caused the price of woodfuel to rise steeply.

This thirteenth century 'wood crisis' in England forced fuel prices up so much that in parts of the country a bundle of firewood cost the same as a bushel of wheat.[245] Price hikes probably hastened a transition to the use of charcoal, which was lighter and therefore easier to transport to the city. But the growing population of the city needed more and more energy to fuel its expansion, and thus pressure of energy demand triggered the change to coal, the price of which had remained stable for decades.

Pressure on the dwindling wood supply in England was eased as a series of famines and then the Black Death struck, finally leaving perhaps a third of the population of Europe dead. It was not until the mid-sixteenth century that London experienced its second 'wood crisis'. Coal consolidated its position in Britain before other European countries because wood shortages emerged sooner there and because of its ability to be transported cheaply on coastal shipping, hence earning the name 'sea-coal'.

Visitors to Britain remarked upon the strange substance burned there. The future Pope Pius II, Enea Sylvio Piccolomini wrote of a 'kind of stone being impregnated with sulphur' in the fifteenth century. Later still, an envoy from Venice named Soranzo described 'a certain kind of earth, almost mineral, which burns like charcoal' which he said would be more widely used but for its bad odour.[246]

In France too, it was the fall of the relative price of wood as against coal that fuelled the industrial revolution from the second decade of the nineteenth century.[247] French coal consumption went from one to fifteen million tons in the half century from 1815, after coal prices had fallen to a level six times cheaper than wood as a source of energy. Adam Smith had written in 1776 that coal, being a 'less agreeable fuel than wood', would need to maintain its price advantage.[248]

The development of the steam engine would change this. Dependent on coal for its operation and widely used in the iron, steel and cotton industries, the 'fire machine' propelled Britain forward to become the supreme manufacturing and commercial power of the early nineteenth century. It was oil and its exploitation, however, which shifted the balance towards the United States in the latter half of the century.

Gas lighting had illuminated factories since the early 1800s. By mid-century it was the most common form of street lighting and where oil was used it was usually whale oil. But with the opening of the Pennsylvania and Ohio oilfields, the days of gas-lamp illumination were numbered. Even the education system was hijacked in the rush to get rich from 'rock oil'. The American Civil War had been a bonanza for railroad barons and businessmen, and they invested their gains heavily in university research and laboratories. The first commercial petroleum refining process came out of just such an investment, the Yale Scientific School. Other laboratories created numerous new consumer goods from the by-products of oil refining, including chewing gum, paraffin and vaseline.[249]

John D Rockefeller started Standard Oil in 1870. Two years later he was refining 10,000 barrels of kerosene a day and by 1904 his company was processing nearly 85 percent of all US crude oil. He drove competitors out of business by fixing transport prices with the railroad companies, developed an effective pipeline system and created markets for lamp oil across the world. To break into China he distributed free oil-lamps in the 'Light of Asia' campaign, thus creating an instant demand for his product.[250] And then came the internal combustion and diesel engines, transforming global energy systems, confirming the primacy of oil as the fuel of the twentieth century and immeasurably increasing man's ability to pollute the atmosphere.

Transports of Delight

The choices involved in personal transport are some of the most important in reducing energy consumption. Ivan Illich[251] wrote almost twenty years ago that

> *the average American male devotes more than 1600 hours a year to his car. He sits in it while it goes and while it stands idling. He parks it and searches for it. He earns the money to put down on it and to meet the monthly instalments. He works to pay for the petrol, tolls, insurance, taxes and tickets. He spends four of his sixteen waking hours on the road or gathering his resources for it.*

There are nearly 500 million motor vehicles in the world today. The exhausts of cars, lorries, buses and motorbikes are spewing a lethal mix of pollutants into our cities and our atmosphere. Sulphur dioxide and nitrogen oxides are the chief pollutants causing acid rain; nitrogen oxides and hydrocarbons combine in sunlight to form the photochemical smog that plagues cities and suburbs from Los Angeles to Jakarta; carbon dioxide is the engine of global warming and carbon monoxide is a threat to the health of city dwellers; the main source of lead pollution is still petrol; asbestos dust from worn brake shoes fills our lungs, and the smoke particles from burning oil carry carcinogens like benz-(a)-pyrene into the deepest recesses of our lungs.

Profligate energy use and pollution may be the two most immediate problems associated with the car, but its rise to primacy in the consumer society has moulded the entire social fabric of the post-industrial age. The history of the rise of the car is immutably intertwined with that of the building of roads. Mick Hamer, who has traced the development of these technological Siamese twins records that cyclists were responsible for starting modern Britain's most powerful industrial interest group, the road lobby. Public opposition to cycling because of the clouds of dust pollution raised on unmetalled roads prompted the two main cycling clubs to start the Roads Improvement Association (RIA) in 1886.

From 1896 the first automobile clubs began to be set up, and they in turn affiliated to the RIA. The number of cars on the road doubled between 1904 and 1905, and the political power of the railway and horse lobbies was already declining. The government was under pressure to reduce the dust pollution which cars were even better than bicycles at producing, and to tackle the problem of road safety.

As Hamer puts it, the RIA offered 'a technical fix, of the kind which has attracted governments ever since. Solving the problem of dust, by improving road surfaces, meant that the government avoided having to confront the issues raised by the spread of the motor car.'[252] In 1907, the year before he became British Prime Minister, Asquith called the motor car 'a luxury which is apt to degenerate into a nuisance'.[253] This must qualify as one of the truly prescient environmental statements of our century.

By 1919, there were plans to build 5000 miles of high quality roads in the UK. In 1932, a new pressure group, the British Road Federation (BRF) was formed and in 1937 there were BRF representatives (and 58 Members of Parliament) in a 255 strong all-expenses-paid delegation to see the new autobahns of the Third Reich. The delegation returned full of praise for the German achievement and recommended that 'the principle of the motorway system be adopted in Britain'. The first motorway opened in 1956 and the first 1000 miles was completed in 1971.

As the motorways took shape, the road lobby turned its attention to cities. Ernest Marples, Minister of Transport from 1959–1964 told a 1963 BRF conference that the motor vehicle was a 'brilliant and beneficial invention' and that 'whether we like it or not...the way we have built our towns is entirely the wrong way for motor traffic'.[254] Growing ownership of motor cars in the 1920s had stimulated 'ribbon development' along the roads that had been surfaced first with John McAdam's chippings, and then with tarmac. The policies of the BRF, Marples and their cronies ensured the destruction of town centres throughout Britain. The final blow was the abolition of Retail Price Maintenance in 1960. This allowed large retailers to undercut the corner shops, and inexorably led to the sterile, car-dependent, out-of-town shopping culture of the 1990s.[255] Britain was not alone in these developments, but it provides a good example of the process which led the car to drive a wedge into the twentieth century and become everybody's favourite personal pollution source.

Numerous studies have shown that no matter how efficient car engines become and no matter how many catalysts are fitted to motor vehicles, it will not be long before the growth in the car population will swamp these technical controls, and air pollution will begin to rise again.[256] While car use in the developed world is still growing slowly, it is in the developing countries, particularly in south-east Asia, that car sales are expanding explosively. Reports compiled by UN/ESCAP show that the number of motor vehicles in Asia increased by almost 25 percent, or by 20 million vehicles between 1984 and 1988, with vehicle ownership more than doubling in both Thailand and Korea during that period.[257] A car rolls off the production line somewhere in the world every second of every day. That's more than 95,000 cars a day – 35 million a year – sold mostly in the western world. In 1990, 15 percent of the world's population bought 79 percent of the world's vehicles.

Some industrial forecasts predict an incredible jump of over 50 percent in worldwide vehicle sales to nearly 75 million in the year 2010. The relatively new Korean market is expected to exhibit the most dynamic growth. Its 300 percent increase in annual sales during the next two decades will lead to there being nearly eight times as many cars on the Korean roads by 2010 as there were in 1990.[258] Pity the poor inhabitants of Seoul. The corollary of such a vast worldwide increase in vehicle sales is that nearly 55 million cars and trucks will have to be scrapped annually. Double the number of 1985, this will present a formidable challenge to the waste and recycling industries.

The motor industry is beginning to respond to environmental pressure from the public and regulation from government. Ford US executive Ernie Savoie wrote in 1990 that the environment is the most serious issue facing his company and that it will affect 'the size and shape of cars, what is in them, how they are made, where they are allowed to go and even who can own them'.[259] Most car manufacturers in Europe and North America now advertise the 'green' features of their cars, such as fuel efficiency or recyclability. Adorning the hoardings are snappy slogans like Volvo's 'green welly' aimed at the guilt-ridden middle classes, Saab's exhortation to get the wind in your hair in its cabriolet, 'just one of the reasons we care about air pollution', or Honda's disarming 'if we can't find a way to live without the car we'd better find a car we can live with'.

Before arriving at a car that can truly be lived with, it is certain that city air will continue to deteriorate into a swirling, choking, aerial swill of smog and smoke, fuelled by ever-worsening traffic congestion. The affliction is spreading fastest in the growing cities of the developing world. Manila, Jakarta, Lagos, Sao Paolo, all are heading for automobile-induced urban coronaries, and their inhabitants towards respiratory grid-lock.

The paradigm of the Third World car crisis is surely presented by Mexico City. In his apocalyptic tale *Christopher Unborn*, Carlos Fuentes described 'The mortal breath of three million motors endlessly vomiting puffs of pure poison, black halitosis, buses, taxis, trucks, and private cars, all contributing their flatulence to the extinction of trees, lungs, throats and eyes.'[260] Crushing winter smogs forced the government to declare a pollution emergency four times between 1988 and 1991, forcibly closing factories and keeping cars off the roads.

It doesn't have to be this way. In the southern Brazilian city of Curitiba, where the population has tripled in the twenty years of his tenure, Mayor Jaime Lerner has guided an astonishing experiment in urban living. He says 'we have simply forseen the problems of rapid urban growth', and in doing so the city planners of this remarkable development may have provided a model for sustainable development.[261]

Strict land-use zoning and growth of the city along predetermined axes have been central to the success, as has the installation of an effective public transport system. Priority lanes allow buses to run separately from cars, and the mass transit system has been made as comfortable and convenient as possible. One inspiration was to solve some of the problems of the slums, or *favelas*, by offering food or bus tickets to their inhabitants in exchange for refuse. This helped to tackle the problem of waste collection in the narrow *favela* tracks, reducing water pollution and pest problems for the people living there, as well as improving resource flow to the poor sectors of the community and enabling them to save money on transport. These and other innovations have led to 1.3 million journeys – in a city of 1.6 million people – being taken on the mass transit system every day, 25 percent lower fuel use than in other Brazilian cities and, above all, a clean and liveable city.[262]

Dominion of the Air

Ivan Illich has some harsh words to say about the airline passenger:[263]

> the occasional chance to spend a few hours strapped into a high-powered seat makes him an accomplice in the distortion of human space, and prompts him to consent to the design of his country's geography around vehicles rather than people.

The importance of mobility in modern society is taken almost to its extreme in the form of air travel. The vast majority of the world's population have never set foot on an aeroplane, and yet this means of transport seems to symbolize the means to escape the humdrum of everyday life for the wealthy consumer class. There is virtually nowhere on the globe that cannot reasonably easily be reached by air. Our world is of a different scale than in the past. The Tolpuddle Martyrs had several hours' walk into the English market town of Dorchester, but the car has cut the journey to no more than 15 minutes. Where it took the Pilgrim Fathers four months to reach New England, the journey now takes seven or eight hours by jumbo jet. The Paris business-woman who can afford *Concord* can boast of her ability to travel to New York for a meeting and return home the same evening.

Today's generation of affluent western children are brought up thinking that flying is as natural a travel choice as walking. Fierce competition between airlines has brought prices for air tickets down substantially in real terms, and as a consequence, passenger numbers are rocketing throughout the world. According to transport analyst Mark Barrett, in the decade up to 1991, the global number of aircraft passengers increased by nearly 50 percent and their average length of journey grew by 10 percent. Demand for air transport is currently growing at 5 percent or more every year, and the total passenger load is predicted to double as early as the year 2010. As air travel takes a strong grip in the burgeoning economies of south-east Asia, passenger demand will double again in the following 20 years.

There is a heavy environmental price to be paid for the freedom of the skies, but apart from campaigns against noise from overflights and development of airports, there has been little public debate about air pollution from aircraft. To remedy this situation, in 1990 WWF commissioned Mark Barrett to write a series of reports outlining the nature of the problem.

This work revealed to a wider audience for the first time that the air industry is one of the fastest growing energy use sectors in the world. Civil aviation alone contributes approximately 3 percent of global carbon emissions and its total emissions are rising by about 4 percent annually. Equally worryingly, aircraft are also responsible for the emission of large quantities of nitrous oxide at or near the border between the troposphere and the stratosphere. Recent research suggests that nitrous oxide emitted at aircraft cruising height of around 12,000 feet may have a substantial impact on global warming.[264] The creation of water vapour at high altitude by aircraft may add another, as yet unquantified, level of greenhouse warming impacts. And the new generation of supersonic aircraft currently in development is likely to increase the threat of ozone depletion as well as adding to global warming.

The impact of aircraft on the greenhouse effect may be to add as much as 5 percent to climate forcing, an inordinately high impact for the relatively few members of the world's population who can actually afford to fly. In 1989, scheduled air services carried about 1.1 billion passengers,[265] probably accounting for no more than about 200 million different individuals. Per capita aviation fuel use in the United States is 154 kg per year, roughly eight times the world or European averages.

But why are people flying? The cut-throat nature of competition in the airline business makes it extremely difficult to obtain detailed passenger information. Most estimates, however, put business travel at around 40 percent in industrialized countries. The one overwhelming factor in the growth of air travel, however, is leisure and tourism, and as such, this sector is critically linked to air pollution by aircraft. In the European Community, tourism expenditure has roughly doubled since the late 1970s, and the number of miles flown increased by 60 percent from 1979 to 1989. Part of this growth is due to holiday-makers' constant search for more exotic destinations. They also want to visit more pristine locations. Plunging into the polluted waters of the Mediterranean is no longer so desirable, so travellers take advantage of cheap flights to 'paradise' islands like Bermuda, Zanzibar or the Maldives to dive among corals and swim in the azure seas. Tourism brings with it both good and evil, but there can be no getting away from the fact that the further you travel by air, the greater are your contributions to the greenhouse effect.

Apart from polluting the air, airlines provide the means to take vast quantities of people to remote places which do not have the infrastructure or capability to cope with additional sewage and even 'cultural pollution'. Airlines can also be directly linked to environmental damage on the ground at the destination. An example of such a relationship can be found in the depressing story of an indigenous tribe in Venezuela.

The Pemone Indians believe the spirit Canaima lives at the top of Auyan Tepuy, the mountain of hell, from which spits forth and tumbles downwards the white foamy streak known to tourists as Angel Falls. Made famous by the American airman Jimmy Angel, who crash-landed near its foot in 1935, the world's tallest waterfall is known by the Pemone as Churun Meru. Its incredible beauty can be approached by flying tight loops in a dead-end canyon, or by boat and foot to the source of the river. Either way, the sight is one of the most spectacular on earth.

The journey to the falls always starts at Canaima, a little settlement whose development demonstrates how great can be the gulf between the interests of tourism and the rights of native people. There are no roads to Canaima and apart from the local headquarters of the hydroelectric power company Edelca, the only place to stay is a 200-bed hotel owned by Venezuela's biggest airline Avensa, which also has exclusive rights to commercial flights into the area. The monopoly would be unremarkable in such a remote place if it wasn't for the fact that this settlement is in the country's largest national park and on the tribal lands of the Pemone people.

The airstrip and the hotel were built using Pemone labour, and the only jobs available to the Indians are in the service of the tour operators, often at below the minimum national wage. Ever since the park was created in 1962 the local indigenous people have been forbidden to cut trees or build on their own land without a permit. Subjected to the will of the National Guard, exploited by the tour company and exposed to the alien cultures of foreigners, the Pemones have lost control of their lives and some have turned to alcohol or prostitution. Many act as guides, cooks and photo subjects for the European and American tourists who flow through Canaima at a rate of more than 40,000 a year.

Julio Moreno Manrique, for instance, lives in a small village close to Canaima where all the six resident families rely entirely on one company, Tiuna, which operates tours from Canaima. Julio guides small groups through the tropical forest, and he and other villagers often rely on leftovers from trekked in chicken barbecues, hosted for visitors in their village. Nearer to the foot of the falls, however, Julio's father and other tribal elders try to maintain the old ways and avoid the pressures to conform with the business plans of Avensa and its two new sister companies, Servivensa and Oturvensa.

For more than 28 years, from 1963, Avensa paid the Venezuelan government only 10,000 bolivar a month for the lease of the land. In 1991, when inflation had corroded this sum to the equivalent of US$160 a month, the lease was increased fifty-fold to US$8000, yet the annual turnover for flights and tourist income for Avensa in Canaima is estimated to be in the region of US$12 million. The Pemones, ancient guardians of Churun Meru, receive nothing except their meagre wages. Even the national parks service, Inparques, gains little. The visitor fee for the park is the same as it was in 1982, 50 bolivar – less than the price of one beer at the Canaima Hotel. The local warden, Jose Chacon, is one of only two Inparques officials supposedly managing the whole of the three million hectare reserve. He relies on Edelca for living quarters and has no motorized transport. His only colleague lives half an hour's flight away in the small town of Ciudad Bolivar and they seldom see each other.

Although for many travellers arriving at the Canaima airstrip aboard an old DC3 in their Banana Republic jungle khakis, the facilities seem less than luxurious, to be a tourist there is a truly privileged experience. The natural spectacle is unrivalled. Throughout the park, more than 100 towering Tepuy mountains shoot steeply, often almost vertically, out of the ancient rock of the Guyana shield between the Amazon and the Orinoco. The tops, sometimes more than 2500 metres above the forest canopy and accessible only by helicopter, are covered in dwarf tropical rainforest, full of endemic species and often shrouded in cloud or mist. Inspiration for Sir Arthur Conan Doyle's novel *The Lost World*, the Tepuys have long been sacred sites for tribes like the Yanomani, Yekuanas, Piaroa and Pemones.

The Pemones aren't alone in experiencing a culture clash

with the outside world, but they are caught in a peculiarly unpleasant kind of tourist trap. Venezuelan environmentalist Julio Cesar Centeno sees it as 'a sad example of how the rights of an indigenous Amazon culture have been sacrificed to benefit mainly a private airline'. Says Centeno: 'The waterfall that comes from Auyun Tepuy is also the attraction for the invaders. In the end the Pemones are the victims instead of the beneficiaries of the beauty of their own land.' Ironically, the Pemones believe that the spirit of Canaima follows them through life and eventually plucks them from it. They must not provoke it and must submit to its will. Can the tourist development be Canaima's will, they wonder?

Energy and Equity

'Energy is inseparable from matter – all material phenomena are associated with energy changes. Energy is also essential to life. And human society cannot survive without a continuous supply of energy.' These are the words of José Goldemberg and his colleagues in *Energy for a Sustainable World* in which they outline the kinds of shifts in energy strategies that will be required to move away from the view that a main indicator of economic growth and human development should be per capita energy consumption. They emphasize the need to focus on the fact that 'energy-use is, after all, only a means of providing illumination, heat, mechanical power, and the other energy services associated with satisfying human needs'.[266]

Energy has, for too long, been associated with development and increased standard of living, and in the modern world, is tied up with the idea of national security. This has led to energy strategies that are driven from the supply side rather than by solid analysis of energy service needs. Per capita consumption of energy can no longer be regarded as an indication of 'modernity', but should be looked at within the context of global development and equity.

At current rates of use, recoverable oil supplies are likely to last only for a further century, yet global annual oil consumption rose from 17 billion barrels in 1970 to 24 billion barrels in 1990. Both natural gas and coal consumption more than doubled over the same period.[267] These fossil fuels are responsible for roughly half of the enhanced greenhouse effect, and carbon emissions from their use rose by a little under 5 percent in the two years from 1987.[268] The lifetime of carbon dioxide in the atmosphere is approximately 120 years, therefore coal burned in 1872, the year Angus Smith first published his theories on acid rain, was still having an impact as the nations at the 1990 World Climate Conference stated that stabilization of CO_2 at 50 percent above pre-industrial levels by the middle of the next century would require 'a continuous worldwide reduction...by 1 to 2 percent per year, starting now'.[269]

The current use of energy in general, and of fossil fuels in particular, is totally unsustainable. Per capita commercial energy consumption in the United States runs at more than twice that of France or even Poland, more than five times that of Mexico or Malaysia, and more than one hundred times that of Somalia, Nepal and Haiti.[270] Developing countries, when asked to take part in the growing plethora of international treaties dealing with global or regional environmental pollution problems, such as the UN Framework Convention on Climate Change, are increasingly asking why they should be penalized in their development when decades of profligate energy use in the industrialized world have brought us where we are today.

It is clear that although historical responsibilty for much pollution lies with the developed world, it would be impossible for all nations to use the same amount of energy per head of their populations as the US currently does for theirs. The new aim should be for the people of all nations to fall within a band of roughly equal per capita energy consumption which provides for an equitable and sustainable quality of life. This must allow for a growth in energy consumption in the developing countries, as energy use is radically reduced in industrialized nations, ultimately leading to convergence at a level which is within the earth's carrying capacity.

But what could this equitable level be? This is not a question that can be answered easily and population growth predictions will be a key factor in this debate. There are more than a quarter of a million people born every day, and the world population is expected to double, to over 11 billion, in just 41 years.[271]

Assumptions about the energy costs of meeting basic needs including food, sanitation, shelter and clothing in the developing countries will need to be combined with programmes aimed at population stabilization and choice in parenthood. Diplomats must decide whether it is possible to assign historical responsibility for past energy use and, perhaps most crucially, the people of the rich countries must take steps towards making changes in their lifestyles.

One study, carried out by energy expert Mark Barrett and concentrating solely on CO_2 emissions, calculated that to achieve the reductions necessary to limit climate change (in an equitable fashion) by the end of next century, most rich nations would have to decrease their per capita carbon emissions by around 90 percent.[272] Although these levels of change are hard to imagine, the consequences of not achieving them are great. For example, if the rest of the world increased their emissions to only 20 percent of current US per capita levels, then carbon emissions would double, causing a 25 percent increase in the rate of global warming.

It is widely believed that changes of a largely technical nature will be able to deliver reductions in energy consumption of between 30 percent and 70 percent in the rich nations over the next few decades. These will involve major improvements in home insulation, efficiency of household appliances and expansion and improvement of public transport systems. Energy efficiency guru Amory Lovins believes 80–90 percent of the $25 billion that Americans spend on air conditioning and cooling their homes, factories and offices could be saved, largely through better design and engineering. Small investments can pay off too. Simply fitting an efficient bathroom showerhead can save more than twice the cost of the new equipment off water and electricity bills in one year.[273]

On top of these technical changes must come lifestyle changes. Inevitably these will include a move away from use of the private car, changes in buying habits and even in choice of clothes. Just as health conscious eating requires a knowledge of the nutritional and calorie content of food, so will environmentally friendly shopping involve making choices informed by an understanding of the energy costs of purchases. Joel Hirschorn and Kirsten Oldenburg have observed that 'pollution prevention is new consumerism. Practising it requires a developed sensitivity to and concern for the life cycle of the thing bought and brought into your home.'[274]

No problem among the many facing our generation, not even the rise and rise of the car, can be regarded as paramount, but the challenge of matching energy and transport services to human development needs and demands for life quality, while simultaneously reducing pollution, will be amongst the greatest. Few doubt that sustainable energy futures are within our grasp, but the political and personal commitment of governments and individual consumers will be sorely tested.

The answers to the energy problem lie not in any single strategy, but in a combination of changes that may take decades to achieve. A shift from supply orientation to demand management; integration of energy and transport planning with each other and with sectors such as environment and education; higher investment in research and development in renewables and behavioural change both in industry and in the home – all will be essential in order to bring about a low-emissions, globally equitable energy system during the twenty-first century.

Chapter Nine

POLLUTION IN THE MAKING: THE EXAMPLE OF ASIA

For nowhere lies beyond man's reach
To mar and burn and flood and leach.
A distant valley is indeed
No sanctuary from his greed.

Vikram Seth

The March of Progress

Roughly half the world's population lives in the 23 percent of the earth's land area that the Asian-Pacific region covers. Seventy percent of the world's agricultural population lives in Asia and some of the worst concentrations of poverty and deprivation outside Sahelian Africa exist in South Asia and Indo-China. Yet many of the fastest growing economies are found in the region and rapid industrialization and urbanization present a growing challenge for an increasing number of countries. The four tigers of Asia – Singapore, Hong Kong, South Korea and Taiwan – have enjoyed phenomenal economic growth over the last quarter decade.

They are now joined by nations including Indonesia, Malaysia and Thailand, and Chinese regions such as Guizhou, Shenzen and Hainan, where industrialization proceeds apace. Thousands of western companies have been attracted to the region by low labour costs. The American sportswear giant Nike, for instance, has 99 percent of its shoes made in Asian countries like Indonesia, where workers in their contract footwear factories earn less than $1.50 a day.[275]

Development in south-east Asia has generally improved literacy standards, reduced birth levels and enhanced health and quality of life among the population at large. Nevertheless, sustained industrialization has had major impacts on the natural resources of the region. Fisheries are under significant pressure, deforestation is increasing and growth in pollution is probably faster than anywhere in the world. In a 1992 survey of 'Asian affluents' more than 60 percent of those questioned said a 'less polluted atmosphere' would most improve their life quality. In the Philippines, Thailand and Taiwan, more than 70 percent of respondents rated pollution their worst problem.[276] The development of environmental legislation and control measures have not kept pace with the rate of emission increases.

Taiwan provides an example of the impacts untrammelled industrial growth is having. With a population density second only to Bangladesh, and 90,000 factories, the small, mountainous and increasingly deforested island is in serious environmental trouble. Ninety-nine percent of sewage is untreated, household garbage rots in open landfills, city smog is worse than Los Angeles and the rivers and coastal waters fizz with industrial effluent. One recent report by Taiwanese scientists warned that 'it is hard to imagine making it to the twenty-first century without a series of environmentally damaging accidents'. Incidents such as acid waste dumping by an ICI subsidiary making plastics, and pollution of fishing grounds by the petrochemical industry have led to the growth of protest movements, massive compensation payments and even anti-pollution riots.[277]

Along with waste and effluent discharges from the burgeoning industrial base in the region, air pollution associated with increased per capita energy use is of serious concern. Energy consumption grew in Asian countries at an average annual rate of 5.3 percent between 1973 and 1987, far outstripping the OECD average of 0.7 percent. Despite this, average per capita energy use in most countries of the region remains at less than 20 percent that of Europe, and below 10 percent of the United States.[278] With greater prosperity, however, comes higher energy use. Singapore already has per capita energy use nearly twice that of Japan, and both South Korea and Hong Kong are fast catching up with the most energy-wasting nations.[279]

The rapid increase in vehicle ownership in many Asian nations has turned cities like Manila, Bangkok and Jakarta into urban hells of smoke, smog and speeding cars. With the possible exception of Malaysia, urban planning in the region has not been able to cope with growth in the car population. In 1983, approximately 70 percent of the vehicles on Indonesian roads were motorcycles, but as the demand for cars grows this percentage is being reduced and both congestion and air pollution grow progessively worse.[280] Car sales are growing by between 7 percent and 8 percent annually in non-Japanese Asia.[281]

Vehicle emissions are responsible for the high levels of lead, carbon monoxide, nitrogen oxides and hydrocarbons in most Asian cities, whilst industry and power generation account for most pollution by sulphur dioxide and particulates. Burning garbage, smoldering underground peat and coal reserves and massive forest fires (now a more frequent occurrence because of logging) can add significantly to the problem. The combination of such concentrations of toxic air pollutants is turning urban air quality into a major health hazard in the region. Lead in the air is slowly being reduced in most western countries after two decades of pressure from health and environmental groups, but the problem is becoming acute in a number of Asian nations.

The Indian cities of Bangalore, Ahmedabad and Calcutta all have high levels of blood lead levels in residents, but it is the more rapidly industrializing countries where the problem is greatest. Chronic lead pollution can damage the nervous system and kidneys of adults, as well as affecting concentration. Children, however, are the most threatened by, and sensitive to, lead contamination. Hyperactivity, aggressiveness and loss of learning ability are three of the most serious threats to exposed children, and brain damage can occur in unborn babies where lead has passed into their blood across the mother's placenta. In Bangkok, where average blood lead levels are roughly 40 percent above the 25 micrograms per decilitre considered dangerous for adults, and seven times higher than the limits for children, babies are born with horrifyingly high lead concentrations in their blood. According to Luaporn Boonakan of Siriraj Hospital, where babies are regularly born with blood lead levels of between 10 and 25 micrograms per decilitre, 'A newborn baby should have a lead level of zero, anything more than that is abnormal.'[282]

But the horror story for Bangkok residents does not end with lead. Sulphur dioxide levels are high, and expected to quintuple over the next twenty years, and levels of the poisonous (but odourless) gas carbon monoxide are already nearly three times higher than maximum allowable safety levels. The chance of getting lung cancer in Bangkok is three times higher than in the provinces.[283]

Acid Rain in Asia

Apart from urban air quality, acid rain is the fastest growing air pollution threat in south-east Asia. By 1987, the Malaysian towns of Petaling Jaya and Senai were suffering rain as acid as that in the UK or West Germany[284] and Hong Kong believes China to be responsible for much of the highly acid rain that falls on the territory.[285]

Usually seen as a problem only for industrialized Europe and North America, acid rain has been quietly eating away at parts of China too for many years. Now the growth of emissions throughout the region is pushing the issue towards the top of the environmental agenda for several other nations. The associated problem of photochemical smog has also been identified as a threat to tropical and sub-tropical forests in India and south-eastern China.[286] The spread of acid rain has already forced the World Bank to set up a major programme to investigate the extent of the threat.

Australian scientists David Roser and Alistair Gilmour have recently completed a review of the situation in Asia, and concluded that the situation appears to be most critical in the north-east Asian region, comprising China, Japan, the Koreas, Taiwan, Hong Kong and Mongolia.[287] These countries account for more than 60 percent of sulphur and nitrogen emissions in Asia. Mongolia and China are in turn receiving very substantial contributions of acid rain from parts of eastern Russia and

Kazakhstan. Indeed, the eastern Siberian region of Russia has a per capita SO_2 emission rate that rivals that of the highly polluted areas of former East Germany. Recent research by Japanese scientists has demonstrated high levels of acidity in rain over the East China Sea near Okinawa.[288]

Acid rain in Asia apears to already be having far-reaching effects, damaging India's Taj Mahal, corroding metal and concrete structures in China and killing rice and wheat crops in some agricultural areas. Unfortunately, despite a growing body of evidence about emissions and deposition, little is yet known about the impacts of acid rain on Asian ecosystems. Damage to forests often occurs as a result of slow acidification of soils, and because acid rain in Asia is a relatively new phenomenon compared with Europe, the soil acidity may not yet have built up sufficiently to cause major visible damage. It seems certain, however, that temperate broadleaved and coniferous forests will eventually react similarly to those in Europe, and there have been some reports of large-scale die-back in the high, ridge-top conifer forests of Bhutan and China's Yunnan province.[289]

The big question mark hangs over the future of tropical ecosystems subjected to acid attack. Most tropical soils are nutrient poor and fairly acid to begin with. The plant species that grow in them have evolved to survive in soils that are within a certain narrow range of acidity. As there is little or no buffering capacity in many tropical soils, even a little bit of acid rain could push these plants beyond their critical limits for survival.

Another feature of wet tropical forest ecosystems that could increase their susceptibility to acid damage is the direct reliance of many species on rainwater. Epiphytic species such as bromeliads and orchids grow on the trunks and branches of trees and gain their water and nutrients direct from the atmosphere through hanging roots. They might be affected first by particularly acid rainfall, as might the animals that rely on pools of water caught between the leaves to breed. Forest frogs and insects play a crucial role in the life-cycle of the forest, often helping to pollinate flowers, and yet many species cannot breed in acid water.[290]

Roser and Gilmour believe that the most threatened ecosystems currently are the subtropical and temperate forests of

113

southern China. Also at risk due to their proximity to rapidly growing sources of acid rain pollution may be the humid and dry tropical forests of parts of the Malay peninsula, Thailand, southwest India and Sri Lanka. From negligible emissions of acid rain at mid-century, Asia is risng to global prominence, producing more SO_2 and NO_x emissions than any other region of the world. Within 20 years these emissions may have doubled.[291]

Dirty Water

And what of the waters of Asia – mystical rivers like the Ganges, the Mekong, the Brahmaputra and the Yangtze? They are all, almost without exception, a sorry mix of open sewer, industrial waste pipe and agricultural drainage ditch. Clogged and suffocated by the constant flow of silt from deforestation in the uplands; starved of oxygen by decaying human excrement; turned sometimes red, sometimes blue, sometimes yellow by an ever-changing mix of toxic chemicals; and filled with pesticide run-off from the green revolution, these rivers are but shadows of their former selves.

Sewage treatment is available for only 2 percent of China's and 4 percent of India's population. In Pakistan the Lyari, Leh and Ravi rivers are severely polluted with municipal waste, as is the Kelani river in Sri Lanka and as are the urban canals of Jakarta. Indeed, it would be difficult to find a single major river running through significantly populated areas in Asia, where sewage is not a problem.[292]

Industrial wastes are a fast-growing threat in the region. The chemical soups that flood from the pipes of tanneries and dyeworks, from pulp and paper factories, from metallurgical works and petrochemical complexes, rubber factories, ore-smelters and mine-workings, palm-oil plants and alcohol distilleries are killing the rivers of Asia. Mining for gemstones is silting up Cambodia's Tonle Sap lake,[293] gold and copper extraction at the huge Freeport and Bouganville mines in New Guinea is killing rivers, smelting operations are poisoning the waters of Cao Hai in China, and the oil industry is about to dip its tainted claws into Vietnam's coastal waters. The booming economies of the region seem not to have time to wait for the enactment or implementation of water pollution laws.

When laws are enacted though, governments are beginning to move against the worst cases of water pollution. After more than $1 million was spent trying to clean industrial waste from the Surabaya River in East Java in 1988, the Indonesian government filed its first suit under new legislation, against a soya bean factory implicated in the pollution. The same week, the Governor of North Sumatra threatened to use the same regulation to prevent a pulp and timber company polluting Lake Toba.[294]

Under most Asian anti-pollution legislation, fines are rarely big enough to deter polluters. A mere US$2000 is the maximum penalty for industrial water pollution in Malaysia, and even in Thailand where the new National Environment Act allows fines up to US$79,000 it is doubtful whether the bureaucracy exists to effectively wield these new policy 'sticks'.[295] The same is true in Taiwan where, since 1992, fines of US$120,000 can be imposed on factories illegally discharging toxic air pollution, but where the Environmental Protection Administration has been kept weak and understaffed by political opponents.[296] In China's Zhejiang province a paltry fine of US$740 for a pharmaceutical plant which had caused US$180,000 of losses to local seafood breeders in just six months, caused angry locals to take the law into their own hands in June 1992 and sabotage the factory to prevent further pollution.[297] The economic costs of inaction are beginning to add up. It is estimated, for instance, that the equivalent of 1 percent of Jakarta's GDP is spent annually just boiling water to render it drinkable.[298]

Pollution monitoring is often years behind that of Europe, Japan or North America. Where the expertise exists, the testing technology often is not available, and where the tests finally do get made, corruption, nepotism, political inertia and ignorance often get in the way of necessary clean-up actions. These latter constraints of course are not solely confined to Asia, but the involvement of men like President Suharto of Indonesia (or in the past, Ferdinand Marcos) at the centre of huge family business empires inevitably makes some major polluters untouchable. If that is not the case, then economics limits effective control measures because, just as in the West, many industries don't see why they should pay for cleaning their effluents when their competitors aren't, and when the chances of detection and prosecution for breaking pollution regulations are practically nil.

There hardly seems to be any activity in the economic furnace of Asia's development that does not cause pollution. When the naturalist Aldo Leopold said 'a man may not care for golf and still be human' he was defending a love of wilderness and nature against the growing popularity of the little white ball.[299] G K Chesterton, on the other hand, believed the game to be akin to slavery in that peer pressure to play kept people from leisure activities they would far rather pursue. Neither Leopold or Chesterton could have imagined the game as an environmental hazard.

Today in Thailand, Korea, Vietnam, Indonesia and Malaysia, environmentalists are hitting out at plans for numerous developments. Japanese and Taiwanese investors plan to build golf courses smack in the middle of two protected tropical forest areas, Vietnam's Thu Duc Forest Park and Thailand's Khao Yai National Park.[300] Thailand expects to have 600 golf courses by the end of the century. The Japanese are Asia's most enthusiastic golfers, and for many it is cheaper to fly to Malaysia or Thailand to play than to join a club in Tokyo. Even in China, where golf was once taboo, memories of the 'Great Helmsman' are put aside as the day of the golf greensman draws nigh. Nine Chinese courses are already in operation and paddies are being drained to create up to 22 more, mainly for customers from Hong Kong and Taiwan.[301]

Apart from destroying valuable agricultural land and natural habitat, golf courses are a significant pollution threat, and the level of fertilizers and herbicides needed to maintain weed-free grass swards in the lush growing conditions of the tropics is phenomenal. Korean citizens groups have protested at the use of persistent toxic chemicals including the organochlorine endosulfan and the organophosphate demeton at golf courses near Seoul. The run-off from many of the courses drains into the catchment area for Paldang reservoir, the source of drinking water for 18 million people.

In Malaysia, the development of a golf resort on the hills of Pulau Redang Island has caused a storm of protest from local

116

environmentalists. Herbicide run-off from the terraced golf course threatens some of the country's most valuable and spectacular coral reefs in the local marine park. Coastal siltation caused by soil erosion during the building of the golf course has already killed some corals, and globally endangered giant clam populations are also threatened, as are coastal mangroves.[302]

The rapid development of golf as a leisure activity in Asia can now be seen as a major pollution threat. A different view is espoused by the Dragon Hill Golf and Country Club in Thailand, whose publicity material suggests 'mother nature's perfection will be joined by inspirations of men in the world of golf, creating a paradise that will be lauded for generations to come'.[303]

The golf boom is just one indicator that the region is changing fast. Asia leads the world in economic growth. The seven strongest nations are maintaining GNP growth at between 5.4 percent and 8.6 percent, with Malaysia logging its fifth year on the trot at more than 8 percent in 1992. The increased level of affluence in the region has stimulated trade between these thriving economies and they are no longer so dependent on the OECD as a market. Thailand, for instance, is expecting to triple telephone ownership to 4.5 million between 1993 and 1997 and Visa International have targeted the region for 30–50 million new Visa cards in the next decade.[304] The hot wind of rapid industrialization is building up to a tumultuous storm of consumption in Asia, but can the environment weather it?

The Chinese Dragon Breathes Fire

Nowhere in Asia is the pool of unsatisfied consumer demand larger than in China. According to the news magazine *Asiaweek*, the country is 'ready for blast-off'. Foreign investment is flooding in to take advantage of cheap labour and an economy stimulated by the fastest growing rate of manufacturing output in the world.[305]

Walk through the maze of market streets of a city like Guangzhou and, mingling with traditional hawkers of deer's antler, bear's paw and rhino horn, live owl, fresh slug and dried lizard, can be found stalls stampeded with buyers for TVs,

Karaoke machines, motorcycle parts and portable telephones. Dutch electrical giant Philips expects a 700 percent increase in its sales of electrical goods over the four year period to 1996[306] and Malaysian automobile company UCM Industrial expects Chinese consumers to be buying a million cars a year by 1995.[307]

The unprecedented increase in Chinese consumption, and of industrial output, could lead to the most massive of pollution problems. Air pollution has already reached appalling levels in many areas. The potential for this trend to continue in the coming years is enormous. The most populous country on earth, with among the lowest per capita consumption of energy and many materials, China could be in for a rough ride. Already, it is one of the few countries outside Europe and North America with a widespread acid rain problem. Photochemical smog is increasing in many Chinese cities. Smoke and dust from industry and domestic coal burning is a health hazard throughout the urban areas, and the level of chemical pollution in some cities would make nineteenth century Manchester seem clean by comparison.

As a result of industrial pollution Chongqing, destined by its geography and climate to experience fogs through much of the year, is now in the nightmarish grip of almost permanent acid fog. Chongqing's emissions of SO_2 are the highest of any Chinese city and, at 900,000 tons annually, are equivalent to a quarter of yearly UK emissions. Thanks to pollution from the burning of high sulphur coal in industry and in the home, the city only experiences 16 days a year without acid fog. Generally more than two full months of dense fog are suffered by the population and respiratory diseases are widespread.[308]

In Shenyang, the largest of the Liaoning cities which are the backbone of one of China's most industrialized regions, the rate of lung cancers among women is one of the highest in the world. The air of Shenyang is thick with smoke from coal burning, and toxic chemicals such as lead, arsenic and fluorine are found in almost every breath of air.

The crippling infant bone disorder, rickets, which has hardly been seen in Europe since the Industrial Revolution, is rife in Shenyang and her sister cities. The lack of sunlight filtering through the grey pall of pollution in heavily polluted parts of

Shenyang and Fushun leave up to 63 percent of children with vitamin D deficiency and the attendant deformities of rickets. A quarter of the children living near the Fushun aluminium smelter also suffer painfully from dental fluorosis. Blood lead levels, as in most polluted Chinese cities, are way above safe levels for children.

Among adults, the combination of their liking for tobacco and exposure to both heavy indoor pollution from coal stoves and chemical, industrial pollution of the outside air gives them a death rate from pulmonary disease five times that of developed countries. Medical scientists in Shenyang have estimated that more than six million working days are lost every year from pollution-related breathing disorders.[309] The use of polluted water for irrigation in Shenyang can cause crops like spinach, rice and cabbage to contain dangerously high levels of toxic metals, including cadmium and chromium. And in Shenyang, along with cities such as Beijing, Xi'an and Jilin, drinking water is often drawn from underground water sources contaminated with arsenic, phenols, cyanide and mercury.[310]

When it comes to acid rain, China has some of the worst problems in the world. Rain with the acidity of battery acid regularly falls in Guiyang, the capital of Guizhou province. Turning rice crops yellow overnight, burning the leaves of vegetation, and corroding cars, metal structures and concrete buildings, the acid downpours remind one of nothing more than the reports of acute industrial pollution in the mill towns of Lancashire 150 years ago. Ironically, Guiyang's acid rain belt, which stretches across south-western China from Yunnan to Hunan, is described in such a large country as 'mainly a local problem'. Yet mountain fog and mist in the rural areas of Mount Fenjin and the conservation area at Cao Hai lake, some 300 kilometres from Guiyang, is often recorded as having a pH of around 3.6, a concentration rarely reached even in central Europe.[311] If a map of the Chinese acid rain affected areas was superimposed on Europe, the problem would be seen to be more than merely local.

Such is the concern over acid rain in China that the government has been carrying out a major research programme since 1984. The World Bank has initiated a project to encourage the

119

transfer of low pollution power station technology to China, and the European Community is coordinating efforts to develop sophisticated computer models which will be able to assess where in China ecosystems and agriculture are most critically threatened by acidification.

The most likely amelioration strategy for China, however, is the substitution of low level power station and factory chimneys for tall stacks. This was precisely the pollution control measure that western nations adopted in the 1950s and 60s in order to try and disperse air pollution away from locally acutely affected areas. Disperse it they did, but in doing so created more trans-boundary air pollution, which, having been pumped high into the atmosphere, was able to travel thousands of kilometres on the winds before falling to earth in other countries as acid rain.

China is likely to save its worst affected areas from further serious affliction for ten years or so, but if the time bought is not used to install desulphurization in power stations, by the end of the the century, acidification will be among China's most widespread and destructive pollution blights.

Lying behind the acid rain problem is the increasing need for energy in China's growing economy. The same issue will make China one of the world's greatest contributors to global warming in a decade or two's time. Already the world's third largest consumer of energy, China's per capita consumption is less than half the world average, and the base level is therefore very low. On current trends, air pollution emissions from energy are liable to quadruple during the next thirty years.

Most Chinese living to the south of the Yangtze River do not have heating in their houses or workplace. So, for instance, in a city like Kunming, a few hundred kilometres from one of China's few tropical forests, occasional freezing temperatures in winter force everybody to don multiple layers of thermal under-wear, thick, down parkas and gloves at work and even at home to sleep in. Chinese people living in these kinds of conditions will soon demand the ability to heat their homes and offices. Many Chinese households currently have only one light bulb, and this too will change.

Government programmes to produce consumables such as refrigerators will increase energy demand as will the spread of motor vehicles, rapidly replacing the bicycle in many cities. Bicycles have already been banned from Shanghai's waterfront so as not to restrict the flow of cars.[312] Since the day in 1958 when Mao Tse-Tung is rumoured to have opened Changchun's new car plant with the words 'At last I'm sitting in a car made by Chinese', Chinese manufacturers have been eclipsed by foreign makers.[313] Most Chinese still cannot afford cars, but many can, and there are now estimated to be more than one million privately owned cars. Beijing alone has over 600,000 vehicles clogging the arteries of the city and competing for space with 8.5 million bicycles.[314] Car manufacture smashed previous records in the first half of 1993, growing by 44 percent to 619,000,[315] with Volkswagen and Citroen being pre-eminent among foreign manufacturers in the country. The giants of the Japanese industry, Nissan, Honda, Mitsubishi and Toyota are now falling over themselves to set up joint ventures.

Clearly the potential for enormous increases in energy-related air pollution exists within China. Hundreds of millions of refrigerators are expected to be manufactured in China in the next decades and the difference between the adoption of low-efficiency models and state-of-the-art technology could be huge. The sale of 700 million fridges (roughly two to three times as efficient as the current ageing UK stock) could mean a saving of around US$7 billion a year off the Chinese electricity bill.[316]

The consequences of rapidly rising Chinese carbon emissions could be catastrophic both for China and the rest of the world. Currently only contributing around 10 percent of global carbon emissions, even with modest improvements in technology and increased efforts towards conservation and efficiency, this figure could quadruple by the year 2030.

Leaving aside the global impacts of increased Chinese carbon emissions, the outlook for the country itself given significant climate change could be gloomy. One study carried out by a team of Chinese and European scientists showed the tremendous potential for global warming to disrupt agriculture and ecosystems in China.

Increases in temperature in the north-western province of Xinjiang, for example, would most likely turn the vegetation from temperate forest and steppe to a warm or subtropical desert.

Alpine vegetation in south-western China would be threatened and northern temperate forest might all but disappear. By 2050 major changes in cropping systems would occur almost everywhere in China. The most significant changes, exacerbated by sea-level rise and flooding, would likely be in the key agricultural area of eastern China.

Reduced water availability over much of China due to global warming would seriously reduce rice and wheat yields, thereby dealing a body blow to the country's efforts to maintain food production on a par with population increase. And while water shortages may be affecting much agriculture inland, sea-level rise could inundate tens of thousands of hectares of alluvial plains on the coast of the South China Sea. This area, home to the oldest continually farmed land in the world (more than 4000 years), currently supports a population of more than 800 people per square kilometre and provides approximately 55 percent of the country's GNP.[317]

These long-term predictions, although worrying to the new generation of environmental scientists and policymakers in China, are taking a back seat to the concern over local pollution and the need for economic development. 'Who's thinking of global warming, when we can barely even breathe' one young Beijinger told the *International Herald Tribune* just before his government signed the Climate Change treaty at the UNCED meeting in Rio. Meanwhile, Jing Wenyong from Qinghua University came closer to the official line when he said 'You can't even talk of economic sacrifice. Above all we must have economic development.'[318]

The kind of economic development many Chinese bureaucrats are talking about involves mega-projects such as the damming of the largest river on earth, the Yangtze. Debated since the nineteenth century, and ostensibly for the purposes of flood control, the Three Gorges Dam would provide more hydroelectric power than any other project in history. It would also displace over a million villagers, and if the project failed, could cause the worst flood ever seen in China.[319] After numerous false starts, protests and petitions, the Three Gorges Dam project was finally approved by a dissension-riven National People's Congress on 4 April 1992.

It is clear that China, with its strong economic growth and the phenomenal success in some regions of a 'free-market with Chinese characteristics' will be a major force in the world of tomorrow. But will it be able to master its environmental and pollution problems to combine an improvement in standard of living for the Chinese people with a marked increase in quality of life? Or is it, as *Time* magazine suggested, merely 'rushing toward the future in a haze of smog and a swirl of poisoned waters'?[320]

Chapter Ten

THE POLITICS OF POLLUTION

> *If I'm not for myself, who will be for*
> *me?*
> *If not this way, How? If not now,*
> *when?*

> Primo Levi

Citizen Power

The Earth Summit was the pinnacle of environmental politics; the culmination of centuries of debate about pollution and habitat destruction. Street children were evicted, armoured cars patrolled the roads and the top states people from more than 150 countries came to Rio. They came to rub shoulders and pontificate on the environment and for a few days there was probably a higher density of despots and democrats, leaders all, than the world has ever seen. They were attended by 9000 journalists and probably 20,000 environmentalists.

There was the official conference, there was the alternative conference, and there was a gathering of indigenous people from all over the world. Two new international treaties were signed – one on biodiversity and one on climate change – and Agenda 21, a 500 page tome outlining what needs to be done to save the planet was agreed. A new Earth Council has been set up, and the UN has created a high-level Commission on Sustainable Development. To quote Pakistan's Prime Minister, Nawaz Sharif, who spoke on behalf of the developing nations, 'After this summit the world will never be the same again.' In some ways he was right.

Before trying to find the road from Rio and predict how it will be travelled, it is necessary to see how the world arrived there in the first place. Previous chapters have examined the development of global environmental and pollution problems; this one looks at the complex politics of dealing with the issues.

Central to the implementation of effective policies to reduce and prevent pollution has been the development of the modern conservation movement, the non-governmental organizations, or as jargon has it, NGOs. Throughout history, individuals or small groups of people have agitated against various types of pollution, but it was the advent of organized conservation societies in the US and Europe that thrust environmental issues to their current heights in political agendas. Set up outside established political parties, these aimed at pressuring whoever was in government to do a better job.

There are tens of thousands of environmental NGOs today, existing in virtually every country where free speech is allowed, and many where it isn't. Their aims and style run the gamut from professional international organizations like Greenpeace and WWF, employing hundreds of staff and working on many and varied issues, through national groups such as the Indonesian Consumer Organization or Canada's Pollution Probe, to tiny local groups campaigning for a single stretch of river or polluting factory to be cleaned up. Many of those working within NGOs do so voluntarily and in their spare time. And in the developing world particularly, NGO activists risk being labelled subversive and being threatened, physically abused or imprisoned.

The NGO movement did not spring up overnight. Herman Melville wrote disparagingly of a 'Society for the Supression of Cruelty to Ganders' in 1851,[321] and pressure groups existed in Victorian England, where the National Smoke Abatement Society campaigned for higher fines for polluters under the Public Health Act, and groups like the Lancashire and Cheshire Association for Controlling the Escape of Noxious Vapours and Fluids, flourished.[322]

While the modern environmental movement is essentially a child of the protest groups of the 1960s, the groundwork was laid in the United States in the last years of the nineteenth century. Just over a hundred years ago, less than thirty years after the end of the American Civil War, and two years after Yosemite and Sequoia became the world's first national parks, John Muir and a group of friends founded the Sierra Club.[323] Its original aims were to lobby for the preservation of 'the forest and other natural features of the Sierra Nevada Mountains', and in 1905 it won the campaign to enlarge Yosemite National Park. In that year too, the Sierra Club started to take an interest in conservation outside California. The last fight of John Muir's life was the failed battle to save California's Hetch Hetchy Valley from flooding by a dam. Stories of this fight inspired the young Californian David Brower, born in Berkeley in 1912, to take up the fight against dams in the American West. Asked in the 1970s why conservationists were 'always against things', Brower replied, 'If you are against something, you are for something. If you are against a dam, you are for a river.'[324]

In Britain, organizations equivalent to the Sierra Club, which would fight for the right to preservation of wild land and access to natural areas, were not formed until well into this century; the Council for the Preservation of Rural England in 1926 and the Ramblers Association in 1935.

By 1959, David Brower was the Sierra Club's controversial Executive Director. When he received a letter from a member which urged him to broaden the Sierra Club's concerns to include tackling the problem of water pollution, he answered: 'The Sierra Club has a passing interest in this, but we still have to be awfully careful about how many new subjects we try to take on until we've digested the old ones.'

Within a few years, recognition of the causes of the 'strange blight' Rachel Carson described in *Silent Spring*, the 1962 publication of which shattered the 'myth of the harmlessness of DDT' and other pesticides, had shaken conservationists from their complacency. Pollution was at last an issue for the NGOs.

As the fall-out from *Silent Spring* continued to have impacts on both sides of the Atlantic, a fledgling conservation group was spreading its wings. The World Wildlife Fund had been incorporated in Switzerland in 1961, and published the Morges Manifesto which predicted that '...senseless human activity [in] the 1960s promise to beat all past records for wiping out the world's wildlife'. The manifesto was signed by 16 prominent conservationists from ten countries, among them Peter Scott, Max Nicholson and Julian Huxley, and included poisoning by toxic chemicals in its catalogue of threats to nature.[325]

In the first report of WWF's progress 'The Launching of a New Ark' in 1965, the International President HRH Prince Bernhard of the Netherlands wrote, 'We are poisoning the air over our cities; we are poisoning the soil itself.... Man is more closely bound up in nature than most people recognise.' Despite these important words, it was more than twenty years before WWF would fully throw itself into the maelstrom of pollution politics.

Back in America, by 1969, David Brower had got the message about pollution but failed in attempts to move the Sierra Club towards a more radical campaigning style. After a very public resignation he set up a new organization called Friends of the Earth (FoE). It was inspired in part by student protests and the work of radical consumer rights activist, Ralph Nader.

The first Director of FoE UK was ex-student leader Graham Searle who went hell-for-leather into campaigns on the fur trade, mining and glass bottles. The bottle campaign was prescient, but a failure. In an inspired media stunt, FoE dumped thousands of used bottles at Schweppes' front door to protest at the use of non-returnable bottles.[326] In its primary purpose, the stunt backfired and manufacturers successfully diverted criticism by promoting bottle recycling. But the FoE campaign to promote the use of returnables had been aimed at saving energy, and hence pollution from power stations. Recycling glass eats up energy.

The bottle dump did put FoE in the headlines, however, and set the pattern for a series of quirky stunts in support of follow-up campaigns throughout the 1970s and early 80s. It also brought potential members flooding in. By 1984, FoE UK had 250 local activist groups and about 30,000 national members, serviced by 17 staff working from a rickety house whose upper floors bent under the weight of papers and filing cabinets.

Gaining international prominence at the same time as FoE was another group started in the heady days of student demonstrations. Developed from the 'Don't Make a Wave Committee' who organized a blockade (protesting against nuclear testing on Amchitka Island) at the border crossing between British Columbia and Washington State in 1969, Greenpeace was born in 1971. The first action was the sailing of a thirty year old halibut seiner, the *Phyllis Cormack* – renamed *Greenpeace* – towards the Amchitka test zone to draw world media and public attention to the nuclear threat.[327]

Greenpeace soon became a household name, and its reputation for boat-based derring-do in 'actions' against nuclear tests and whaling, combined with an uncanny ability to provide the media with great stories and better pictures earned them a dedicated following amongst young environmental idealists.

Through the 1970s and early 80s, WWF, FoE and Greenpeace developed in different ways. While WWF developed an impressive record in developing countries, working in the field on practical conservation, both FoE and Greenpeace opted for a more confrontational approach involving demonstrations, actions, preparation of scientific evidence for government enquiries and hearings, and unrelenting use of the media.

They both also put pollution at the top of their agendas. Greenpeace initially highlighted marine pollution and toxic industrial waste while FoE took on pesticides and freshwater pollution. By the mid-1980s, both organizations had active programmes on energy and transport policy and on acid rain. By this time of course, there were already thousands of other NGOs around the world working in similar ways, and their power was beginning to be felt.

Danish NGOs won a complete ban on throwaway beverage packaging;[328] community campaigns after the Love Canal tragedy in the US led to the endowment of the SuperFund; Australian NGOs won concessions on mining in National Parks; phosphates were banned in detergents in Switzerland; and a thousand other victories won.

And where did the activists of these pressure groups learn their trade? Well they learned mainly from experience, and from each other. The chronicler of Greenpeace's early days, Robert Hunter, credited the Quakers for the new organization's non-violent witness approach and Marshall McLuhan for their media savvy. Dave Brower drew his inspiration from the radical writings of the wilderness advocates John Muir and Aldo Leopold, and from his friend the photographer Ansel Adams.

Des Wilson, the campaigner who, having emigrated from New Zealand to Britain, launched the housing charity Shelter, the Campaign for Lead Free Air (CLEAR) and the Freedom of Information Campaign says he learned his tradecraft from the battle over prohibition in his dry, home town of Oamaru in the 1950s. Later he took up the 'citizen action' battle cry of Ralph Nader.[329] By and large, for all the anti-pollution campaigners it was a process of trial and error. Some campaigns were won and others lost, but the lessons were always observed and remembered. As the 1980s drew to a close and the early 90s began to pass, and at a time when more people worldwide belong to environmental NGOs than ever, pollution activists have had to begin to rethink their campaign techniques.

A string of pollution disasters – pesticides at Bhopal; radioactivity at Chernobyl; chemicals from Sandoz; and oil in Alaska and during the Gulf War – have heightened public sensitivity to the issues. This enhanced public interest, and a rat-a-tat succession of ever more complex global atmospheric pollution problems hitting the headlines during the last decade – from acid rain and smog, through ozone depletion to the greenhouse effect – have spawned a new generation of environmental specialists in newspapers and on radio and TV. These correspondents often know as much or more about the issues than the fresh-faced young environmentalists trying to brief them, and can't be easily manipulated to write or broadcast stories with the correct 'spin'. Until the mid-1980s environmentalism in the developed world was an unfashionable occupation, often low-paid and relying on volunteer helpers. It ran on a kind of swashbuckling romanticism and was populated by a crazy mixture of dreamers, back-to-the-landers, scientists and computer whizzes. You would have needed a Pantone book to describe the different shades of 'green' that were represented.

The FoE veteran Charles Secrett claimed to have got his job as whaling campaigner on the strength of having written his college dissertation on the novels of Herman Melville. And Greenpeace's current Campaigns Director, Chris Rose, once likened the growth of the movement in Europe during the early 1980s to the birth of a religion. First came the prophets, people like Ralph Nader, David Brower, Rachel Carson, Rofe Pomerance, Des Wilson, Jonathon Porritt and Petra Kelly, then the disciples (campaigners like Rose himself and Secrett), and then finally the acolytes and followers. These latter were less questioning in their adherence to pressure group dogma; they came not so much out of idealism, but because they had read of the environmental campaigns in style magazines, or seen dazzling actions on TV. Some of them came over the wall from industry, and most of them wanted a better wage, training and most of all, job security.

Eric Hoffer once wrote 'When a mass movement begins to attract people who are interested in their individual careers, it is a sign that it has passed its vigorous stage; that it is no longer engaged in molding a new world but in possessing and preserving the present.'[330] That crunch is coming for the campaign groups.

In parallel developments, the public has grown tired of picture-grabbing actions such as riding around in Zodiac inflatables or hanging banners from smoking chimney stacks, a global recession has hit charity fund-raising, and industrial lobbies such as those for coal, nuclear power and agrochemicals have stepped up their efforts to counter pollution arguments. All of this makes it harder for the advocacy groups.

Humorist P J O'Rourke caught a popular public sentiment when he wrote 'ecological problems won't be solved by special interest groups spreading pop hysteria and merchandising fashionable panic' and he questioned their motives by saying 'cures for environmental problems might even endanger the environmentalist fad'.[331]

The environmental groups have always been in their element when on the attack. They have brought new threats to the fore of public debate, they have been able to scare the public into caring about such esoteric subjects as stratospheric chemistry, nuclear physics and biotechnology. Excellent on problems, the environmentalists frequently struggle on solutions. So much so that industry appears to have taken the initiative in many areas. Shell Oil has a Non-traditional Business Division researching alternative fuels and forestry problems, Freeport Mines and Chevron Oil claim to be helping develop sustainable development projects in New Guinea; the masters of the disposable watch, Swatch, are pioneering solar vehicle technology and computer giant Bull has set up the Arbor programme to protect forests. Worldwide, industry probably now employs tens of thousands of environmental specialists.

Look in any issue of *Time* or *Newsweek* and you can expect to see several companies extolling their efforts to make the world a better place and prevent pollution. Saab, Volvo, Honda, Nissan, Asea Brown Boveri, Honeywell, Sanyo, ICI, Evian, DuPont: all regularly buy space to show how green they are. Long in direct opposition to industry, NGOs are having to explore a new relationship. One which allows them to remain critical, while recognizing that without the support of industry the urgent reductions that are needed in emissions of virtually every chemical pollutant are unlikely to be achieved.

The dilemma is a tough one for campaigners, for as soon as they get off the backs of industry, the pressure for change is gone. On the other hand, if they continue to uncritically criticize, they will eventually lose much credibility in the eyes of their supporters. The way out of this impasse is to maintain diversity among environmental groups (with some staying more aloof from dialogue with industry than others), keep questioning industry on its green commitment and start to offer more solutions. All of this can be underpinned by increased attention to legislation, both national and international. Happily, all the signs so far point towards a strong and flexible response by the NGO movement to the challenges of the post-UNCED world.

There Ought to be a Law Against it!

The issue of legislation is crucial. Businesses are unlikely to radically change their operations, especially where cost is involved, if competitors do not also have to do so. It's the 'level playing field syndrome', and once everyone is on the same footing industry will act. International trade can act as a spoiler in this regard because national governments are often reluctant to pass legislation that will put their country's businesses at a disadvantage against those of others. It can also militate against the developing world when multinational companies transfer their most polluting operations out of a highly legislated industrialized country to a nation with a need for foreign investment, cheap labour costs and lax pollution standards.

Although pollution legislation had appeared sporadically in the past (for instance Britain's Alkali and Public Health Acts of the nineteenth century, and the fog-driven UK Clean Air Act of 1956), it was not until the late 1960s and early 1970s that most OECD countries began to develop comprehensive pollution legislation. The US Clean Air Act (1970) and West Germany's Federal Emission Control Act (1974) were typical of these new pollution laws.

To go with the regulations came regulatory bodies: The Swedish National Environmental Protection Board (1969); the US Environment Protection Agency (1970); and the French Ministry for the Protection of Nature and the Environment (1971), for instance. Most countries then went on to produce specific laws to control air pollution and protect surface waters, and a little later to regulate toxic chemicals and require environmental impact assessments.[332]

By the end of the 1980s it was clear that regulations to prevent emissions to individual media – air, water, soil – could never be totally effective as environmental problems would just be transferred from one place to another. The acid rain debate in the UK provided a good example of the problem. Britain had a pollution policy of using the best practicable means (BPM) for emissions control. The policy of building high stacks to disperse air pollution from coal-fired power stations had been adopted in the 1960s, but merely resulted in increased acid rain in neighbouring European countries. A policy described by the German environmental analyst Helmut Weidner, as 'pollution export' or 'problem-shifting as best practicable means'.[333]

The alternative sought by British environmentalists was to fit flue gas desulphurization equipment to power station chimney flues. This could reduce SO_2 emissions by 80 percent, but it required the digging of limestone from National Parks, and the disposal of thousands of tons of gypsum sludge afterwards. Promotion of this policy brought both Friends of the Earth and

Greenpeace into conflict with another powerful NGO, the Council for the Protection of Rural England (CPRE) in the late 1980s. Britain's SO_2 emissions are still the same as they were in 1985[334] and no resolution to the acid rain problem has yet been attempted.

Problems of this sort, however, gave rise to the concept of Integrated Pollution Control (IPC). The idea, which is enshrined in 1990 Environmental Protection Acts for both Britain and Sweden, is that the environmental impacts of all pollution control measures should be assessed and that the administrative and legislative structure in government should be integrated. Discharge permitting and licensing should be centralized under one authority to avoid the problem-shifting of the past. Needless to say, IPC has not yet proved effective in either country in its practical application.

As recognition of the need for national pollution laws has grown, so too has the trend towards international agreements and treaties. It was this trend more than any other that persuaded environmentalists to start to discard their Arran sweaters and replace them with collars, ties and even suits. These days, there is a breed of pollution activist who is happiest in the subterranean corridors and windowless meeting rooms of the UN negotiating circuit. Many have law degrees and some have even been on the other side of the fence, as civil servants and diplomats in previous occupations. They mix with the transformed old-style campaigners and a growing contingent of radical environmentalists from southern countries such as Bangladesh, Kenya, Venezuela, Nigeria and Malaysia. These campaigners may spend weeks away from home as the travelling roadshow of international environmental negotiations moves from Geneva to New York to Nairobi and back again, dealing this week with ozone, the next with toxic waste and the next with climate change.

The days can be long. Final agreement on the IPCC's first report on the science of climate change only came at 3.30 in the morning of the last day after a week of tense negotiations in Sweden in 1990. NGOs were instrumental at that meeting in persuading Brazil to remove its objection to the text (campaigners explained that the exhausted international press corps waiting in the foyer were drafting stories making Brazil the villain of the piece). Claude Martin, Director General of WWF, himself now desk-bound in Switzerland after many years in the forests of India and West Africa, has dubbed this new breed of environmentalist 'airport biologists' because they spend more time in Schipol, Dulles and Heathrow than they do in the habitats they are trying to save.

But international pollution law is becoming increasingly important, and NGOs are consolidating their ability to influence it. The *Wall Street Journal* profiled Tessa Robertson, then WWF-UK's Pollution and Energy Officer, in the lead up to UNCED, saying 'very few people have developed her level of direct access to the top negotiators in the European Community.'[335]

International conservation treaties have been around for some time. The International Convention for the Regulation of Whaling was signed in 1946, and the International Plant Protection Convention in 1951. Water pollution began to be dealt with by the Council of Europe's 1968 'Agreement on the Restriction of Certain Detergents in Washing and Cleaning Products' requiring said products to be '80 percent biodegradeable'. 1972 saw the signing of the London Dumping Convention and 1973 the Convention for the Prevention of Pollution from Ships. These were followed by a string of regional seas conventions aimed at protecting marine environments such as the Mediterranean and the Caribbean.[336]

Many of these agreements were long in the negotiation stages, and flawed when it came to implementation. The real breakthroughs in multinational law came with the air pollution agreements. The first of these was the 1979 Geneva Convention on Long-range Transboundary Air Pollution, under which protocols agreeing reductions of SO_2, NO_x and VOCs have since been signed. This convention is pioneering the use of the 'critical loads' concept for pollution control, and is a focus for NGO activity on acid rain.

After the discovery of the hole in the ozone layer, and against the initial opposition of powerful industrial interests including DuPont, ICI and Atochem, the Vienna Convention on Protection of the Ozone Layer was signed in March 1985. Just 30 months later, in September 1987 after an unprecedentedly short negotiation period for such a complex and controversial issue, the Montreal Protocol on Substances that Deplete the Ozone Layer was completed. Richard Benedick, chief US negotiator on the Protocol said afterwards, 'perhaps the most extraordinary aspect of the treaty was its impositon of substantial short-term economic costs to protect human health and the environment against unproved future dangers – dangers that rested on scientific theories rather than on firm data.'[337]

The ozone negotiations took place on the cutting edge of science, with a massive space-age research programme producing, it seemed, new results almost daily. Ultimately the politicians had to choose between the pleas of industry and the growing body of still inconclusive scientific evidence that chlorofluorocarbons (CFCs) were destroying the earth's protective ozone layer. That they chose science was in no small measure due to the NGOs.

CFCs had long been a political issue in the US. In 1974 two California-based researchers, Mario Molina and Sherry Rowland, suggested that these chemicals could damage stratospheric ozone, and that they could persist as pollutants in the atmosphere for decades and even centuries. By 1978, the use of CFCs as propellants in aerosol spray cans had been banned in the US. CFCs were still used for a variety of other purposes including refrigeration, foam-blowing and cleaning electrical circuitry, and although the aerosol ban temporarily cooled public ardour for action in the US, the scientific research continued.

The announcement of the discovery of the Antarctic ozone hole in 1985 caused an eruption of NGO activity in the US and a strong flurry of international environmental diplomacy. But the NGOs of Europe seemed to slumber through the debate. While ICI and Atochem were virtually writing British and French negotiating positions on the international treaty, environmentalists in those countries were struggling with the political aftermath of the Chernobyl disaster, and with the threat of worsening acid rain damage.

Two US NGOs in particular changed this. The World Resources Institute and Natural Resources Defence Council sent two of their most persuasive advocates, Irving Mintzer and David Wirth, to Europe to explain to NGOs including Britain's

FoE and Germany's Bund, that Europe was the crucial block in the debate, its chemical industry was running the show, and now was the time to act. This occasioned some debate within European NGOs. At FoE, for instance, some people argued that ozone depletion was far too obscure a problem to ever expect the public to understand.

Nevertheless, FoE was first off the blocks in the UK, and with a constant flow of up-to-date scientific and political information coming from Washington, they were rapidly able to develop a campaign. Taking advantage of the fact that the aerosol industry published its own guide to propellants, they published *The Aerosol Connection* in 1986. It was a comprehensive list of those aerosols that did not use CFCs, and urged consumers to buy only products from this list. Journalist Geoffrey Lean wrote an exclusive on the new campaign in the *Sunday Observer*, and within two weeks FoE had had thousands of requests for the information.

They followed up with a plea to consumers to hand back the CFC-blown foam clam-shells their hamburgers came in – a plea which resulted in incredible interest from popular press and radio, and led within a few days to McDonalds announcing it would be switching to non-CFC foams. By this time Greenpeace had also entered the fray, and even Prince Charles had accidentally got himself (and the issue) on the front page of the tabloids when he was quoted as saying Lady Diana should stop using hair sprays.

The speed with which public opposition to CFCs grew across Europe weakened the chemical companies' cosy lobbying position, and governments gradually came around to the US government's view that there was sufficient evidence available to justify phasing out CFCs completely. The Montreal Protocol had been a groundbreaking piece of legislation for several reasons. For the NGOs, it showed them, perhaps for the first time, the power of international environmental lobbying. They had been observers before, but had either been marginalized, or effective only through the efforts of a few dedicated individuals. On the ozone treaty though, the NGOs were instrumental in the success of its rapid negotiation.

From a legal standpoint there were several key innovations:

the Protocol pioneered the concept of a fund to support the transfer to developing countries of less polluting technologies; it agreed to phase out a group of industrial chemicals before they were conclusively proved to have caused damage; and most importantly it allowed for a strengthening of control measures every four years. The parties to the Montreal Protocol met again in Copenhagen in November 1992 to do just that.

The ozone talks proved to have been a valuable rehearsal for the much more politically charged and complex negotiations on climate change that began with the formation of the IPCC in 1988. Whereas CFCs were only ever a small sector of the chemicals industry, and just six countries produced approximately 95 percent of them, the climate debate centred primarily around emissions from fossil fuels. This was an issue that governments really had a stake in, and the fossil-fuel and road lobbies from Washington to Tokyo went into overdrive to ensure they wouldn't end up like the CFC manufacturers had done. Behind them was ranged the support of oil dependent nations such as Saudi Arabia, Kuwait and Venezuela.

The scientific evidence for climate change is compelling, however, and the Montreal Protocol had already established that strong measures could be agreed before final proof showed that global warming had occurred. It is likely that without the Montreal Protocol, the climate treaty could not have been ready in time for signing at Rio. But the negotiations did succeed, and a treaty was signed, so where will the road from Rio take us?

Towards the Next Millenium

In the world after the Earth Summit, it is no longer possible to say that environment and development problems are being ignored, but neither is there any evidence that they are close to solution. The Rio meeting took environment to the top of the international political agenda for a brief moment, but government follow-up was negligible in the succeeding two years. The dimensions of the global environmental problem are now clearer than at any time in history, public awareness of the magnitude of the challenge before humankind is growing and governments are at last trying to come to grips with the issues. To translate understanding into action, however, is likely to prove a Herculean task.

The problem of pollution cannot be treated in isolation from the generalized environmental crisis. Neither can it be separated from its Siamese twin, consumption, itself an integral part of the population debate. The three issues together lay bare the raw nerve of equity among people and whichever one you plan to deal with matters little as all are irrevocably connected through the web of trade, development and debt and stuck together with the glue of international economics.

The demonstration of such a mire of linkages between complex issues, making their collective solution seem still more complex should not, however, engender feelings of hopelessness. For, just as an apparently simple issue cannot effectively be addressed without understanding its root causes and ramifications, neither can a multi-faceted problem be satisfactorily resolved without detailed attention to its constituent facets. In short, we must deal with the lone smoking chimney, the single chemical dump and individual lifestyles, with as much energy and ingenuity as can be brought to bear, just as we must with the social and economic causes of the wider problem.

This approach requires that individuals, whether they be in industry, government, agriculture, home-making or any other field, begin to take greater responsibility for the impacts of their actions. To do this, they must be better informed about these impacts, better enabled to make choices between products or actions, and most of all, committed to an ethic of environment and equity.

Central to a world in which personal commitment to change can help achieve concrete improvements in quality of life and reduce pollution will be political, economic and technological change. There is likely to be no more powerful catalyst than internalization of environmental costs in economic calculations. This would help to move industry towards cleaner production, governments to fairer trade and consumers to be more discriminating.

Apart from full cost accounting, the reduction of pollution will require radical action in three specific areas: increased emphasis on energy efficiency and the rapid introduction and diffusion of renewable sources; massive reduction of levels of per capita resource consumption in the industrialized world; and a new development path for the Third World, in which long-term environmental security is not sacrificed at the altar of short-term economic growth.

There are indications that some businesses are at last beginning to recognize the future role they will have to play in planetary maintenance. It is unlikely that any solution to local or international pollution problems can be found without the cooperation of the private sector, and increasingly governments and environmentalists are looking to industry for partnerships in breaking through traditional practice. In the field of technology transfer for instance, initiatives to bring energy efficient light-bulbs to Mexico, low-emissions motorcycles to India and clean coal technology to China have all begun to meet with some success in the last few years.

Indications of the potential role that private businesses could play in developing new technologies and work practices that contribute towards sustainable development can be seen already in the US. For example, Carl Weinberg, Head of Research and Development for California's Pacific Gas and Electric (PG&E) company has argued for a new value system based on 'enoughness'. Weinberg believes that this is different from sustainability or efficiency, and defines 'enoughness' as 'meeting a need for getting the job done in a manner using the least resources and having the least environmental impacts. It means considering all the options to meet a need, and adopting the option that is just enough.' In the electricity business, this paradigm requires utilities to invest widely in energy efficiency and conservation measures rather than just trying to sell as much electricity as possible. PG&E already has more than 600,000 residential customers participating in energy efficiency programmes, and in 1992 they saved enough energy between them to power 90,000 homes.[338] Similar initiatives are taking off with electricity companies across the US.

Smaller organizations also have a role to play in bringing a new outlook to business activities. Based in Denver, Colorado, the World Engineering Partnership for Sustainable Development has taken as its mission breaking down the barriers between specialists in the engineering profession and promoting integrated planning. One project they have been involved in is the planning for a low-impact industrial complex in Taiwan, where waste products from one factory become the raw materials for another, and where waste heat is also utilized down the line. In Chicago, too, the search for new solutions is hotting up. The Center for Neighbourhood Technology (CNT) has been working to help polluting industries modernize and clean up, where environmental groups might just have watched those companies go out of business. Their efforts for the electro-plating industry in Chicago and Los Angeles have encouraged companies in those cities to clean up and increase jobs, while the metal-finishing industries have been laying people off every-where else in the US. Now CNT plans to do the same for the printing and dry-cleaning industries, and they are already working to promote a non-solvent based cleaning system first developed in the UK.[339]

In the OECD countries, many companies are supporting the reduction of motor vehicle pollution by allowing staff to work from home part of the time, or by encouraging car-pooling, public transport and bicycling. In 1993, several large US compa-nies, including Rockwell, Bell Atlantic and GTE Telephone Operations founded the National Telecommuting and Telework Association. The objective of the organization is to provide advice to companies which want to start programmes in their offices, and also to work with policy-makers to create a 'telecommuting friendly' marketplace.[340] One telecommuting programme initiated around traffic-snagged Seattle by the Washington State Energy Office in 1990 has already proved hugely successful. Working with 280 people in 25 organizations, the scheme demonstrated that people who telecommuted one day a week typically drove more than 1200 miles less a year, and saved nearly 50 gallons of petrol. According to project manager Dee Christensen, 'we estimated that if 15 per cent of the workforce could telecommute just one or two days each week it would reduce energy consumption of gasoline in the Puget Sound area by about 14 million gallons a year.' The Seattle experiment, and others throughout the USA, have also shown that telecommuting can substantially increase job performance and productivity.[341] With more than two million people telecommuting in the USA already, this could turn out to be the wave of the future, with benefits for employer, employee and environment. McDonalds, 3M, Norsk Hydro and Volvo are among many that have launched huge pollution reduction programmes in their factories and outlets. But even with the growing variety and success of such initiatives, there is a limit to the possibilities for fundamental change. A clean car still requires construction materials and roads. A clean pesticides factory still makes pesticides.

In *Beyond the Limits*,[342] Donella and Dennis Meadows and Jorgen Randers tried to define what it was that society had to change to start to make these moves towards a sustainable world:

> *They are the cultural norms and practices that maldistribute income and wealth, that make people see themselves primarily as consumers and producers, that associate social status with material accumulation, and that define human goals in terms of getting more rather than having enough.*

In theory elected governments do what the voters want, businesses exist to serve their stakeholders and meet consumer demand, and institutions like the World Bank provide the financial and technical resources for Third World development. In reality, governments generally direct their programmes at staying in power, businesses create demand for products and promote a slavish struggle up the ladder of consumption, and the Bank's prime objective is a good rate of financial return. Until such attitudes and practices change, pollution will continue to eat away at and rot the fabric of our societies.

If humankind is to enter the next millenium with a modicum of hope that the carrying capacity of the earth will not be overwhelmed within a few generations, then environment and sustainable development must be lifted to the very top of the international agenda and maintained there. In the twilight of the twentieth century we must tackle these problems with the fervour that Victorian reformers devoted to sanitation and education. Environmentalists hoped that the election of President

Clinton to the most powerful job in the world in 1992 was a move in the right direction.

With Clinton came Vice President Al Gore, who once asked if God chose 'an appropriate technology when he gave human beings dominion over the earth?'[343] Also elevated into the Clinton cabinet was Governor Bruce Babbitt of Arizona. These two politicians were the vanguard of a hopeful new breed of Democrat environmentalists lifted into power. Babbitt was in optimistic mood in the Spring of 1993 when he told National Park Service employees, 'It's a time to...think big, to dream, to have a sense of urgency.'[344] But so far the sense of urgency has eluded the American people and good intentions have not been sufficient to change the measure of the nation. The logging and grazing disputes of the West remain unresolved, agricultural intensification continues to eat away at soils and habitat, and not a single pollution problem is on the road to resolution. Debates that many had hoped would lead to partnership towards environmental and economic gain between business leaders and the government increasingly smack of old-fashioned compromise under pressure by the Clinton administration.

In America as elsewhere, however, there are some rays of hope. President Clinton's US National Climate Action Plan has at least turned to face the greenhouse gas problem head on, even if it has not yet fully provided the means to effectively cut emissions. The Super-Fund for the clean-up of toxic sites is being revised to become more effective, and the new Clean Water Bill has taken on the issue of banning chlorine. Meanwhile, the devastating earthquake that hit Los Angeles and destroyed key parts of the city's road network in January 1994 may have at least one positive consequence. The political aftermath of the shock may force a rethink in transport planning that will see roads given second place to light rail in the rebuilding programme.

These advances are small indeed when viewed in the light of the global deterioration of natural habitats, resource depletion and the inexorable build-up of pollutants in the environment, but they are at least advances. Each tiny change for the better in the way we manage our environment, each successful project and every new initiative aimed at reducing human impacts may act as a beacon to guide others. Eventually, with common under-standing of the unsustainability of modern fossil-fuelled economies and consumption patterns, we may learn to confront our collective self-destructive urge and through individual action as well as community determination, step back from the brink.

The pollution crisis that we have been heading towards since humankind began its long march through civilization is finally reaching its peak. A small alteration in the course of human history could help to bring an end to the pollution problem. But to slow the momentum of human affairs which is carrying us towards our own nemesis will require an unprecedented global effort. As UNCED's Secretary General, Maurice Strong, said at the end of the Rio meeting, 'after millions of words, the last two – let's begin'.[345]

NOTES AND REFERENCES

1 Janssens, Paul, *Palaeopathology: Diseases and Injuries of Prehistoric Man* (London: John Baker, 1970)
2 Brimblecombe, Peter, *The Big Smoke* (London: Routledge, 1988)
3 Janssens, Paul, *Palaeopathology: Diseases and Injuries of Prehistoric Man*, op cit
4 Hammond, Norman, 'Lead Poisoning Blamed for Rome's Fall' in *The Times*, London, 1 January 1994.
5 Ponting, Clive, *A Green History of the World* (London: Penguin Books, 1991)
6 Braudel, Fernand, *The Identity of France – Volume One: History and Environment* (London: Fontana Press, 1988)
7 From the abridged *Johnson's Dictionary*, 7th Edition, printed and published by W Strachan in 1783, and the Times Books facsimile folio edition *A Dictionary of the English Language* of 1979
8 Mellanby, Kenneth, *The Biology of Pollution* (London: Edward Arnold, 1982)
9 Ponting, Clive, *A Green History of the World*, op cit
10 Meybeck, Michel et al, *Global Freshwater Quality: A First Assessment* (Oxford: Basil Blackwell, 1989)
11 Brimblecombe, Peter, *The Big Smoke*, op cit
12 Ziegler, Philip, *The Black Death* (London: Penguin Books, 1975)
13 Brimblecombe, Peter, *The Big Smoke*, op cit
14 Ziegler, Philip, *The Black Death*, op cit
15 Fitter, R S R, *London's Natural History* (London: Collins, 1946)
16 Ziegler, Philip, *The Black Death*, op cit
17 Ponting, Clive, *A Green History of the World*, op cit
18 Braudel, Fernand, *The Identity of France – Volume Two: People and Production* (London: Fontana Press, 1991)
19 Quoted in Brimblecombe, Peter, *The Big Smoke*, op cit
20 Quoted in Fitter, R S R, *London's Natural History* op cit
21 McNeill, William, *Plagues and People* (London: Penguin Books, 1979)
22 Suyin, Han, *The Crippled Tree* (London: Triad Grafton, 1989)
23 Hughes, Robert, *The Fatal Shore* (London: Pan Books, 1988)
24 McNeill, William, *Plagues and People*, op cit
25 Wohl, Anthony, *Endangered Lives: Public Health in Victorian Britain* (London: Methuen, 1984)

26 ibid
27 Williams, Gwyn, *When was Wales* (London: Penguin Books, 1985)
28 Hughes, Gareth, *A Llanelli Chronicle* (Llanelli: Llanelli Borough Council, 1984)
29 Fitter, R S R, *London's Natural History*, op cit
30 Brimblecombe, Peter, *The Big Smoke*, op cit
31 ibid
32 ibid
33 Entry for 24–30 January 1684 in Evelyn, John, *The Diary of John Evelyn* (London: Oxford University Press, 1959)
34 White, Gilbert, *The Natural History and Antiquities of Selborne* (London: Nathaniel Cooke, 1853)
35 Dennis, Richard, *English Industrial Cities of the Nineteenth Century* (Cambridge: Cambridge University Press, 1986)
36 Quoted in Wohl, Anthony, *Endangered Lives: Public Health in Victorian Britain* op cit
37 Sinclair, Upton, *The Jungle* (London: Penguin Books, 1986)
38 Fitter, R S R, *London's Natural History*, op cit
39 Rose, Chris, *The Dirty Man of Europe: The Great British Pollution Scandal* (London: Simon and Schuster, 1990)
40 Harrison, Paul, *The Third Revolution* (London: Icarus, 1993)
41 Braudel, Fernand, *Civilization and Capitalism: 15th–18th Century (Volume 1), The Structures of Everyday Life* (London: Fontana Press, 1985)
42 Mumford, Lewis, *The City in History: Its origins, its transformations and its prospects* (London: Penguin Books, 1961)
43 Porter, Roy, *English Society in the Eighteenth Century* (London: Penguin Books, 1982)
44 Hobhouse, Henry, *Seeds of Change: Five plants that transformed mankind* (London: Papermac, 1992)
45 Porter, Roy, *English Society in the Eighteenth Century*, op cit
46 ibid
47 Briggs, Asa, *Victorian Cities* (London: Penguin Books, 1990)
48 Engels, Friedrich, *The Condition of the Working Class in England* (London: Penguin Books, 1987)
49 'The Thames', *The Illustrated London News* (p 112), 30 July 1859
50 'Main Drainage of the Metropolis', *The Illustrated London News* (p 203), 27 August 1859
51 ibid
52 Quoted (p 523) in Mumford, Lewis, *The City in History*, op cit
53 ibid
54 Briggs, Asa, *Victorian Cities*, op cit
55 ibid
56 Cronon, William, *Nature's Metropolis: Chicago and the Great West* (New York: Norton, 1992)
57 ibid

58 ibid
59 Mumford, Lewis, *The City in History*, op cit
60 ibid
61 'After Cleaning Up Its Air, Los Angeles Faces a Smoggy Future', in *The New York Times*, 26 November 1993
62 World Health Organization and United Nations Environment Programme, *Urban Air Pollution in Megacities of the World* (Oxford: Blackwell, 1992)
63 Quoted in Marshall, Peter, *Nature's Web: An Exploration of Ecological Thinking* (London: Simon and Schuster, 1992)
64 Braudel, Fernand, *The Structures of Everyday Life*, op cit
65 Lindop, Grevel, 'The Countenance of Nature: Towards an aesthetics of every thing, every place, every event', in *The Times Literary Supplement*, no 4720, 17 September 1993)
66 ibid
67 Wordsworth, William, 'Ode Composed Upon an Evening of Extraordinary Splendor and Beauty', in Gill, Stephen (ed), *The Oxford Authors: William Wordsworth* (Oxford: OUP, 1990).
68 Clare, John in *The Oxford Authors: John Clare* (Oxford: OUP, 1984)
69 Thomas, Keith, *Man and the Natural World: Changing attitudes in England 1500–1800* (London, Penguin Books, 1984)
70 Marshall, Peter, *Nature's Web*, op cit
71 Thoreau, Henry David, *Walden and Civil Disobedience* (New York: Viking Penguin, 1986)
72 ibid
73 Quoted in Shabecoff, Philip, *A Fierce Green Fire: The American Environmental Movement* (New York: Hill and Wang, 1993)
74 Bronowski, Jacob, *William Blake and the Age of Revolution* (London: Routledge and Kegan Paul, 1972)
75 Quoted in Campbell, Colin *The Romantic Ethic and the Spirit of Modern Consumerism* (Oxford: Basil Blackwell, 1990)
76 Pepper, David, *The Roots of Modern Environmentalism* (London: Routledge, 1990)
77 ibid
78 Quoted in Marshall, Peter, *Nature's Web*, op cit
79 ibid
80 Unpublished notes circulated at the Second Morelia Symposium, Mexico, January 1994
81 Quoted in Thomas, Keith, *Man and the Natural World*, op cit
82 White, Gilbert, *The Natural History and Antiquities of Selborne*, op cit
83 Clare, John, in *The Oxford Authors: John Clare*, op cit
84 Defoe, Daniel, *A Tour Through the Whole Island of Great Britain* (London: Penguin Books, 1971)
85 Quoted in Purseglove, Jeremy, *Taming the Flood: A History and*

Natural History of Rivers and Wetlands (Oxford: OUP, 1989)

86 Thoreau, Henry David, *Walden and Civil Disobedience*, op cit

87 Russell, Bertrand, *Authority and the Individual* (London: Unwin 1990)

88 ibid

89 Abbey, Edward, *Desert Solitaire: A Season in the Wilderness* (New York: Ballantine Books, 1971)

90 From author's notes of a speech made at the Second Morelia Symposium, Mexico, January 1994

91 Smith, Adam, *The Wealth of Nations* (London: Penguin Books, 1986)

92 Cooke, Alistair, *America* (London: BBC, 1973)

93 Hemming, John, *The Conquest of the Incas* (London: Macmillan, 1970)

94 Hammond, Norman, 'Lead poisoning blamed for Rome's fall', op cit

95 Quoted in Young, John, *Mining the Earth* (Washington DC: Worldwatch Institute, 1992)

96 Hemming, John, *The Conquest of the Incas*, op cit

97 Mellanby, Kenneth, *The Biology of Pollution*, op cit

98 *International Marketing Data and Statistics 1992* (London: Euromonitor, 1992)

99 Boorstin Daniel, *The Americans: The National Experience* (London: Sphere Books, 1988)

100 Robertson, Patrick, *The Shell Book of Firsts* (London: Ebury Press, 1974)

101 Thoreau, Henry David, *Walden and Civil Disobedience* op cit

102 Galbraith, J K, *The Affluent Society* (London: Penguin Books, 1991)

103 Goodman, Jordan and Honeyman, Katrina, *Gainful Pursuits: The making of Industrial Europe 1600–1914* (London: Edward Arnold, 1992)

104 Briggs, Asa, *Victorian Things* (London: Penguin Books, 1988)

105 Raban, Jonathan, 'New World', in *Granta 29*, Winter 1989

106 Most of the information on the chronology of inventions has been garnered from Boorstin Daniel, *The Americans: The Democratic Experience* (London: Sphere Books,1973)

107 Boorstin Daniel, *The Americans: The National Experience*, op cit

108 Goodman, Jordan and Honeyman, Katrina, *Gainful Pursuits: The Making of Industrial Europe 1600–1914*, op cit

109 Wohl, Anthony, *Endangered Lives: Public Health in Victorian Britain,* op cit

110 Goodman, Jordan and Honeyman, Katrina, *Gainful Pursuits: The Making of Industrial Europe 1600–1914*, op cit

111 Gorz, André, *Ecology as Politics* (Montreal: Black Rose Press, 1980)

112 Boorstin Daniel, *The Americans: The National Experience*, op cit
113 Ponting, Clive, *A Green History of the World*, op cit
114 Rybczynski, Witold, 'Waiting for the Weekend', in *The Atlantic*, vol 268, no 2, August 1991
115 Gorz, André, *The Politics of Ecology*, op cit
116 Quoted in Boorstin Daniel, *The Americans: The Democratic Experience*, op cit
117 Munro, David and Holdgate, Martin (eds) *Caring for the Earth: A Strategy for Sustainable Living* (London: Earthscan, 1991)
118 Quoted in Macrae, David, *The Americans at Home* (Glasgow: Horn and Connell, 1885)
119 Dompka, Victoria, *Population and the Environment: A WWF Discussion Paper* (Gland, Switzerland: WWF-International, 1994)
120 ibid
121 Braudel, Fernand, *The Structures of Everyday Life*, op cit
122 Rural Planning Department, *Nakuru: District Development Plan 1989–1993* (Nairobi: Ministry of Planning and National Development, 1989)
123 *The Illustrated London News* (p 561), 8 June 1872
124 'Population and Progress', *The Illustrated London News* (p 485), 19 November 1859
125 Smith, Adam, *The Wealth of Nations*, op cit
126 Clarke, Robin, *Water: The International Crisis* (London: Earthscan,1991)
127 Meybeck, Michel et al, *Global Freshwater Quality: A First Assessment*, op cit
128 *Our Planet, Our Health* (Geneva: World Health Organization, 1992)
129 ibid
130 'Health and Sanitation in India', in *Nature*, vol 122, no 3081, 17 November 1928
131 *Our Planet, Our Health*, op cit
132 Anderson, John Ward, 'Turtles Going to Work at Cleaning the Ganges', in the *International Herald Tribune*, 26 September 1992
133 Ayres, Ed, 'Whitewash: Pursuing the Truth About Paper', in *Worldwatch*, vol 5, no 5, September/October 1992
134 Carson, Rachel, *The Sea Around Us* (London: Panther Books, 1969)
135 Conrad, Joseph, *The Mirror of the Sea & A Personal Record* (Oxford: Oxford University Press, 1988)
136 Spence, Jonathan, *The Search for Modern China* (London: Century Hutchinson, 1990)
137 Pernetta, John and Elder, Danny, *Cross-sectoral, Integrated Coastal Area Planning (CICAP): Guidelines and Principles for Coastal Area Development* (Gland, Switzerland: IUCN, 1993)
138 Hinrichsen, Don, *Our Common Seas: Coasts in Crisis* (London:

Earthscan, 1990)

139 *The Environmental Programme for the Mediterranean* (Washington DC: The World Bank, 1990)

140 Grove, Richard, 'Origins of Western Environmentalism', in *Scientific American*, July 1992

141 Ozanne, Julian, 'Expanding Tourism Presents Dilemmas', in the *Financial Times*, 14 September 1992

142 *Annual Tourism Statistical Report 1990* (Kuala Lumpur: Tourism Development Corporation Malaysia, 1991)

143 Wells, Sue, *Coral Reefs: Valuable but Vulnerable* (Gland, Switzerland: WWF-International, 1992)

144 *The State of the Marine Environment* (Nairobi: Joint Group of Experts on the Scientific Aspects of Marine Pollution (GESAMP), UNEP, 1990)

145 *The Environmental Programme for the Mediterranean*, op cit

146 Rose, Chris, *The Dirty Man of Europe: The Great British Pollution Scandal*, op cit

147 Gourlay, K A, *Poisoners of the Seas* (London: Zed Books, 1988)

148 *World Resources 1990–91* (Oxford: Oxford University Press, 1990)

149 Houghton, J T et al (eds), *Climate Change 1992: The Supplementary Report to the IPCC Scientific Assessment* (Cambridge: Cambridge University Press, 1992)

150 *Gulf War Environmental Damage Assessment.* Unpublished First Progress Report (Project 4623) from IUCN to WWF-International, July 1992

151 *World Resources 1990–91*, op cit

152 Keckes, Stjepan et al, *Policies for Marine Conservation and Resource Management*, unpublished report to WWF-International and IUCN, 1992

153 Jansen, Ronald et al, *Environmentally-friendly Packaging in the Future: A scenario for 2001* (Amsterdam: Vereniging Milieudefensie, 1991)

154 *World Resources 1990–91*, op cit

155 *World Resources 1992–93*, (Oxford: World Resources Institute, Oxford University Press, 1992)

156 *Our Planet, Our Health*, op cit

157 Hurst, Peter, *Chemicals Control Policy in the European Community* (Gland, Switzerland: WWF-International, 1992)

158 'Polluter Still Paying for Minimata Poisonings', in *Industry and Environment*, vol 15, no 1–2, January–June 1992.

159 Colborn, Theodora et al, *Great Lakes, Great Legacy?* (Washington DC: The Conservation Foundation, 1990)

160 Begley, Sharon and Glick, Daniel, 'The Estrogen Compex' in *Newsweek*, 21 March 1994

161 Colborn, Theodora, vom Saal, Frederick, and Soto, Ana, 'Developmental Effects of Endocrine-Disrupting Chemicals in

Wildlife and Humans' in *Environmental Health Perspectives*
vol 101, no 5, October 1993
162 ibid
163 Pearson, Charles (ed), *Multinational Corporations, Environment,
and the Third World* (Durham, North Carolina: Duke University
Press, 1987)
164 ibid
165 Selcraig, Bruce, 'Bad Chemistry', in *Harper's*, vol 284, no 1703,
April 1992
166 ibid
167 DeLillo, Don, *White Noise* (New York: Penguin Books, 1986)
168 Galbraith, J K, *The Affluent Society*, op cit
169 Quoted in Dennis, Richard, *English Industrial Cities of the
Nineteenth Century*, op cit
170 ibid
171 Wohl, Anthony, *Endangered Lives: Public Health in Victorian
Britain,* op cit
172 Brimblecombe, Peter, *The Big Smoke*, op cit
173 'Muriatic Acid Gas', in *The Illustrated London News* (p 266),
16 March 1872
174 Park, Chris, *Acid Rain: Rhetoric and Reality* (London: Methuen,
1987)
175 Rose, Chris, *The Dirty Man of Europe: The Great British Pollution
Scandal*, op cit
176 Isakkson, Folke, 'Air Attack', speech to environmentalists and
writers at the Morelia Symposium 'Hacia el fin del milenio',
Mexico, 1991
177 *Digest of Environmental Protection and Water Statistics, 1991*
(London: Department of the Environment, HMSO, 1992)
178 Meybeck, Michel, *Global Freshwater Quality: A First Assessment*,
op cit
179 *Critical and Target Loads Maps for Freshwaters in Great Britain*,
unpublished third report to the Department of the Environment by
the UK Critical Loads Advisory Group, March 1992
180 Houghton, John et al (eds), *Climate Change: The IPCC Scientific
Assessment* (Cambridge: Cambridge University Press, 1990)
181 Houghton, John et al (eds) *Climate Change 1992: The
Supplementary Report to the IPCC Scientific Assessment*, op cit
182 Parry, Martin *Climate Change and World Agriculture* (London:
Earthscan, 1990)
183 'The Deadly Hitch-hikers', in *The Economist*, 31 October, 1992
184 Williams, H E and Bunkley-Williams, L, 'The World-wide Coral
Bleaching Cycle and Related Sources of Coral Mortality', in *Atoll
Research Bulletin* (Washington DC: The Smithsonian Institute,
1990)
185 Ellison, J C and Stoddart, R D, 'Mangrove Ecosystem Collapse

During Predicted Sea-level Rise: Holocene Analogues and Implications' in *Journal of Coastal Research*, vol 7, no 1, 1991; and Ellison, J C *Hungry Bay Mangrove Swamp, Bermuda: Present Condition and Future Management* (Bermuda: Bermuda Biological Station for Research, 1991)

186 Rose, Chris and Hurst, Philip, *Can Nature Survive Global Warming?* (Gland, Swizerland: WWF-International, 1992)

187 Markham, Adam, Dudley, Nigel and Stolton, Sue, *Some Like it Hot: Climate Change, Biodiversity and the Survival of Species* (Gland, Switzerland: WWF-International, 1993)

188 ibid

189 ibid

190 *United Nations Framework Convention on Climate Change: Text* (Geneva: UNEP/WMO, 1993)

191 'First step towards tighter curbs on greenhouse gases' in *The ENDS Report*, no 229, (London: Environmental Data Services, February 1994)

192 Clinton, William, and Gore, Albert, *The Climate Action Plan*; (Washington, DC: The White House, 1993)

193 Milliken, Robert, *No Conceivable Injury* (Sydney: Penguin Books Australia, 1986)

194 ibid

195 ibid

196 Lawrence, Christopher, *Cellular Radiobiology* (London: Edward Arnold Limited, 1971)

197 Milliken, Robert, *No Conceivable Injury*, op cit

198 Fishlock, David, 'The Healing Machine that Radiated Death', in the *Financial Times*, 29 September 1988

199 Keepin Bill, 'Nuclear Power and Global Warming' in Leggett, Jeremy (ed), *Global Warming: The Greenpeace Report* (Oxford: Oxford University Press, 1990)

200 Vidal, Gore, 'Through the Film Screen, Darkly' in the *International Herald Tribune*, 4 September, 1992

201 Durning, Alan, *How Much is Enough?* (London: Earthscan, 1992)

202 ibid

203 Paine, Thomas, *Rights of Man* (London: Penguin Books, 1985)

204 Ekins, Paul, 'The Sustainable Consumer Society: A Contradiction in Terms?', in *International Environmental Affairs*, Fall 1991

205 Bahro, Rudolf, *From Red to Green* (London: Verso, 1984)

206 Galbraith, J K, *The Affluent Society*, op cit

207 George, Mathews, 'Seeing the Green Light', in *South*, September 1989

208 Galbraith, J K, *The Affluent Society*, op cit

209 *European Marketing Data and Statistics* (London: Euromonitor, 1992)

210   Meadows, Donella et al, *Beyond the Limits: Global Collapse or a Sustainable Society?* (London: Earthscan, 1992)
211   Clarke, Robin *Water: The International Crisis*, op cit
212   Orwell, George, *Coming up for Air* (London: Penguin Books, 1979)
213   Steinbeck, John, *The Grapes of Wrath* (London: Penguin Books, 1978)
214   Durning, Alan, *How Much is Enough?*, op cit
215   'Food for Thought', in *The Economist*, 29 August 1992
216   Durning, Alan, *How Much is Enough?*, op cit
217   Rifkin Jeremy, 'Anatomy of a Cheeseburger', in *Granta* 38, Winter 1991.
218   Morse, Laurie, 'McDonald's Puts Beef into its Expansion', in the *Financial Times*, 11 November 1992
219   *International Marketing Data and Statistics 1992* (London: Euromonitor, 1992)
220   Kranendonk, S, and Bringezu, S, 'Major Material Flows Associated with Orange Juice Consumption in Germany', *Fresenius Environmental Bulletin 2*: 455–460 (Basel: Birkhauser Verlag, 1993)
221   Prendergrast, Mark, *For God, Country and Coca-Cola: The Unauthorized History of the World's Most Popular Soft Drink* (London: Weidenfeld and Nicolson, 1993)
222   Young, John, 'Aluminium's Real Tab', in *Worldwatch*, vol 5, no 2, March 1992
223   Pearce, Fred, *The Dammed: Rivers, Dams and the World Water Crisis* (London: The Bodley Head)
224   Quoted in Hirschorn, Joel and Oldenburg, Kirstin *Prosperity Without Pollution* (New York: Van Nostrand Reinhold, 1991)
225   Bongaerts, Jan, *'The German Packaging Ordinance'*, in *Environmental Business Law Review*, July 1992
226   Jansen, Ronald et al, *Environmentally-friendly Packaging in the Future: A Scenario for 2001*, op cit
227   *Market Research Europe*, January 1992
228   *Market Research GB*, February 1992
229   *Supermarket Business*, September 1991
230   *Nikkweek*, 21 September, 1991
231   Mayer, Caroline, 'Millions Are Turning Off the Tap And Buying by the Bottle', in *The Washington Post*, 2 February 1994
232   *Jakarta Post*, 25 August 1992
233   Elsworth, Steve (ed), *The Environmental Impacts of the Car* (Amsterdam: Greenpeace International, 1991)
234   ibid
235   Galbraith, J K, *The Affluent Society*, op cit
236   Schneider, William, 'The Suburban Century Begins', in *Atlantic Monthly*, July 1992
237   Knight, Stephen, *The Selling of the Australian Mind: From First*

     Fleet to Third Mercedes (Sydney: William Heinemann Australia, 1990)

238 Ekins, Paul, 'The Sustainable Consumer Society: A Contradiction in Terms?' op cit

239 Rybczynski, Witold, 'Waiting for the Weekend', op cit

240 Ekins, Paul, 'The Sustainable Consumer Society: A Contradiction in Terms?', op cit

241 Pearce, Fred, *The Dammed*, op cit

242 Meybeck, Michel et al, *Global Freshwater Quality: A First Assessment*, op cit

243 Balagot, Beta, 'Environmental Impacts of Geothermal Energy Systems', in Ramani, K V et al (eds) *Burning Questions: Environmental Limits to Energy Growth in Asian-Pacific Countries during the 1990s* (Kuala Lumpur: Asian and Pacific Development Centre/WWF-International, 1992)

244 Debeir, Jean Claude et al, *In the Servitude of Power* (London: Zed Books, 1991)

245 ibid

246 Gimpel, Jean, *The Medieval Machine* (London: Pimlico, 1992)

247 Braudel, Fernand, *The Identity of France – Volume Two: People and Production*, op cit

248 Smith, Adam, *The Wealth of Nations*, op cit

249 Collier, Peter and Horowitz, David, *The Rockefellers: An American Dynasty* (New York: Signet, 1977)

250 Boorstin Daniel, *The Americans: The Democratic Experience*, op cit

251 Illich, Ivan, *Energy and Equity* (London: Marion Boyars Publishers, 1974)

252 Hamer, Mick, *Wheels Within Wheels* (London: Routledge & Kegan Paul, 1987)

253 Quoted in a review of 'The Problem of Motor Transport: an Economic Analysis', in *Nature*, no 3117, vol 124, 27 July 1929

254 Hamer, Mick, *Wheels Within Wheels*, op cit

255 Sherlock, Harley, *Cities are Good for Us* (London: Paladin 1991)

256 See for example Fergusson, Malcolm et al, *Atmospheric Emissions from the Use of Transport in the United Kingdom: Volume One – The Estimation of Current and Future Emissions* (Godalming, UK: WWF, 1989)

257 *State of the Environment in Asia and the Pacific 1990* (Bangkok: United Nations Economic and Social Commission for Asia and the Pacific, 1990)

258 Done, Kevin 'Prepare for Rising Demand in Less Developed Countries', in the *Financial Times*, 11 September 1991

259 Leadbeater, Charles, 'Environmental Issues will Transform Car Industry says Ford Report', in the *Financial Times*, 22 October 1990

260 Fuentes, Carlos, *Christopher Unborn* (London: Picador, 1990)
261 Musson, Peter, 'Civic Deals on Wheels' in the *Geographical Magazine*, June 1992
262 ibid
263 Illich, Ivan, *Energy and Equity*, op cit
264 Barrett, Mark, *Aircraft Pollution* (Gland, Switzerland: WWF-International, 1991)
265 ibid
266 Goldemberg, Jose et al, *Energy for a Sustainable World* (New Delhi: Wiley Eastern Limited, 1988)
267 Meadows, Donella et al, *Beyond the Limits: Global Collapse or a Sustainable Future?*, op cit
268 Houghton, John et al, *Climate Change 1992: The Supplementary Report to the IPCC Scientific Assessment*, op cit
269 'Conference Statement' from the Second World Climate Conference, Switzerland, 7 November 1990
270 *World Resources 1990–91*, op cit
271 Dompka, Victoria, *Population and the Environment*, op cit
272 Barrett, Mark, *Energy, Carbon Dioxide and Consumer Choice*, unpublished report to WWF-International
273 'No Sweat: Stay Cool and Save Billions', in *Rocky Mountain Institute Newsletter*, vol 8, no 2, Snowmass, Colorado
274 Hirschorn, Joel and Oldenburg, Kirsten, *Prosperity Without Pollution*, op cit
275 Clifford, Mark, 'Spring in Their Step', in *Far Eastern Economic Review*, 5 November 1992
276 'Asia Lifestyles', in *Far Eastern Economic Review*, 10 September 1992
277 Maitland, Alison, 'Anger Grows Over Taiwan's Polluted Success Story', in the *Financial Times*, 13 October 1989
278 Markham, A and Ramani, K V, 'Environmental Challenges for Asian-Pacific Energy Systems in the 1990s' in *Energy and Environment*, vol 2, no 4, 1991
279 Schwarz, Adam, 'Looking Back at Rio', in *Far Eastern Economic Review*, 28 October 1993
280 Leinbach, Thomas and Sien, Chia Lin *South-East Asian Transport* (Singapore: Oxford University Press, 1989)
281 Done, Kevin 'A Chequered Pattern', in the *Financial Times*, 20 October 1992
282 Walsh, Michael, '1991:The Year in Review', in *Car Lines*, vol 9, no 1, January 1992
283 ibid
284 Peng, Leong Chow et al, *Report on Rain Acidity Analysis Based on Data from the National Acid Rain Monitoring Network* (Kuala Lumpur: Malaysian Meteorological Service, 1988)
285 *Air Quality in Hong Kong* (Hong Kong: Hong Kong Environmental

Protection Agency, 1985)
286 Kickert, Ronald, and Krupar, Sagar, 'Forest Responses to
     Tropospheric Ozone and Global Climate Change: An Analysis' in
     Environmental Pollution (London: Elsevier, 1990)
287 Roser, David and Gilmour, Alistair, *Acid Rain in Asia and the
     Pacific: Its Current Extent and Repercussions for Biological
     Conservation in Asia and the Pacific* (in Press, WWF-International,
     Switzerland)
288 Walsh, Michael, 'Japan Reports High Acid Rain Levels in East
     China Sea', in *Car Lines*, vol 10, no 5, October 1993
289 John MacKinnon, personal communication
290 McDowell, W H, 'Potential Effects of Acid Deposition on Tropical
     Terrestrial Ecosystems', in Rodhe, Henning and Herrera, Rafael
     (eds) *Acidification in Tropical Countries: SCOPE 36* (Chichester:
     John Wiley and Sons, 1988)
291 Roser, David and Gilmour, Alistair, *Acid Rain in Asia and the
     Pacific*, op cit
292 *State of the Environment in Asia and the Pacific 1990*, op cit
293 Schwarz, Adam, 'Looking Back at Rio', op cit
294 Brown, John Murray, 'Indonesia Takes Action Against Polluter', in
     the *Financial Times*, 4 November 1988
295 Vatikiotis, Michael, 'Plenty of laws but Little Action' and Handley,
     Paul, 'New Rules but Old Attitudes', both in *Far Eastern Economic
     Review*, 29 October 1992
296 Wehrfritz, George, 'Asia's Richest but also Dirtiest', in *Far
     Eastern Economic Review*, 29 October 1992
297 Hendry, Sandy, 'Policy Reform When it Suits', in *Far Eastern
     Economic Review*, 29 October 1992
298 Schwarz, Adam, 'Looking Back at Rio', op cit
299 Leopold, Aldo, *A Sand County Almanac* (New York: Ballantine
     Books, 1989)
300 Hiebert, Murray, 'Green Fees', in the *Far Eastern Economic
     Review*, 20 August, 1992 and Patrick Corrigan in the letters column
     of the *Bangkok Post*, 29 January 1992
301 Karp, Jonathan, 'Greens for the Reds', in *Far Eastern Economic
     Review*, 29 October 1992
302 Unpublished material from WWF Malaysia, prepared for a press
     conference in Kuala Lumpur on 20 April 1992
303 'Golf Courses Destroying Land for Agriculture and Resources', in
     the *Bangkok Post*, 18 October 1991
304 Taber, George, 'Growing, growing...', in *Time Magazine*,
     14 Septenber 1992
305 'Ready for Blast-off', in *Asiaweek*, 10 July 1992
306 'The Big Picture', in *Asiaweek*, 24 September 1992
307 'The Newest Money Rush', in *Asiaweek*, 10 July 1992
308 Jianbo, Zhou et al, 'Chongqing Fog and its Health Effects', in

press, WWF/Chinese Academy of Meteorological Sciences

309 Zhaoyi, Xu, 'The Health Effects of Air Pollution on Citizens in Liaoning Cities', in press, WWF/ Chinese Academy of Meteorological Sciences

310 Smil, Vaclav, *The Bad Earth: Environmental Degradation in China* (London: Zed Press, 1984)

311 Jiling, Xiong et al, 'Sources and Effects of Air Pollutants in Guizhou Province', in press, WWF/Chinese Academy of Meteorological Sciences

312 Maddox, Bronwen, 'It Will get Worse', in the *Financial Times*, 18 November 1993

313 Thomson, Robert, 'More Joint Ventures are Sought', in the *Financial Times*, 18 November 1993

314 Associated Press, 28 April 1993

315 Associated Press, 4 September 1993

316 Barrett, Mark, *Carbon Emissions from Fuel Burning in China: Patterns and Prospects* (Gland, Switzerland: WWF-International, 1991)

317 Hulme, Mike et al, *Climate Change due to the Greenhouse Effect and its Implications for China* (Gland, Swizerland: WWF-International, 1992)

318 WuDann, Sheryl, 'China is Gaining Fast in CO_2: Energy Gets Priority Over Environment', in the *International Herald Tribune*, 26 May 1992

319 Pearce, Fred, *The Dammed*, op cit

320 Burton, Sandra, 'The East is Black', in *Time Magazine*, 29 April 1991

321 Melville, Herman, *Moby Dick or The Whale* (London: Chancellor Press, 1985)

322 Wohl, Anthony, *Endangered Lives: Public Health in Victorian Britain,* op cit

323 'A Centennial Celebration: 1892–1992', in *Sierra*, vol 77, no 3, May/June 1992

324 Quoted in McPhee, John, *Encounters with the Archdruid* (New York: The Noonday Press, 1990)

325 Scott, Peter (ed), *The Launching of a New Ark* (London: World Wildlife Fund and Collins, 1965)

326 Pearce, Fred, *Green Warriors* (London: The Bodley Head, 1991)

327 Hunter, Robert, *The Greenpeace Chronicle* (London: Picador, 1980)

328 Durning, Alan, *How Much is Enough?*, op cit

329 Wilson, Des, *Pressure: The A to Z of Campaigning in Britain* (London: Heinemann, 1984)

330 Quoted in Steele, Shelby, 'The New Sovereignty', in *Harper's*, vol 258, no 1706, July 1992

331 O'Rourke, P J, *Parliament of Whores* (New York: Vintage Books,

1992)

332 Weale, Albert, *The New Politics of Pollution* (Manchester: Manchester University Press, 1992)

333 Weidner, Helmut, *Clean Air Policy in Britain* (Berlin: Edition Sigma, 1987)

334 *Digest of Environmental Protection and Water Statistics* no 14 (Norwich, UK: Government Statistical Service, HMSO, 1992)

335 Allen, Frank, 'Earth Movers: Five People to Watch at the Rio Conference', in the *Wall Street Journal*, 1 June 1992

336 Sand, Peter, *Lessons Learned in Global Environmental Governance* (Washington DC: World Resources Institute, 1990)

337 Benedick, Richard, *Ozone Diplomacy* (Cambridge, Massachusetts: Harvard University Press, 1991)

338 'From Rio to the Capitols', Conference Proceedings from State Strategies for Sustainable Development, Commonwealth of Kentucky, 1993

339 ibid

340 'Telecommuting catching on as a way to save energy, cut pollution' in *Quad Report*, vol 1, no 11, November 1993

341 'From Rio to the Capitols', op cit

342 Meadows, Donella et al, *Beyond the Limits: Global Collapse or a Sustainable Future?*, op cit

343 Gore, Albert, 'What is Wrong with Us?', in *Time Magazine*, 2 January 1989

344 Quoted in Kenworthy, Tom, 'The Lord of the Land', in *The Washington Post*, 23 January, 1994

345 *Terra Viva*, no 12, Rio de Janeiro, 15 June 1992

INDEX

Index compiled by John Tooke